轻豪宅风范 II

LIGHT MANSION STYLE II

奢而不侈 豪而不夸

LUXURIOUS YET NOT EXTRAVAGANT,
GORGEOUS YET NOT EXAGGERATED

深圳视界文化传播有限公司 编

中国林业出版社
China Forestry Publishing House

序言

P R E F A C E

DO EXPRESSIVE DESIGNS TO TOUCH YOURSELF

做会表达的设计，感动自己

SHEN ZHEN DK INTERIOR DESIGN CO., LTD　Shuren Xu
深圳市帝凯室内设计有限公司　徐树仁

Secluding in the mountains and living in a thatched cottage is a kind of lifestyle loved by ancient Chinese. However in modern society with rapid development of science and technology, those who can live such a lifestyle are less and less while many people choose a favorite residence in a corner of the city or build a commercial place. This makes interior design more important in the process of development.

Born in 1970s, I'm glad to witness the large-scale constructions, the rapid development of real estate and the increasingly speed upgrading of urbanization process in the era. The significant improvement of people's life quality has an obvious influence on interior design. The rapid development of design industry promotes the entire society and improves life quality, which plays an important role. Interior design also has an extraordinary significance in the survival spaces of the entire human society. Without architecture, there is no design. Luxurious five-star hotels on two sides of streets, shining advanced markets, theme restaurants with elegant tone and mansions hidden in the city reconditely present modern people's pursuit and yearning of high-quality life.

Interior design has experienced numerous changes in China for decades, develops and matures gradually and forms the design style defined by countless people. However when it becomes more mature, the immature itself is found. All so-called style are constantly defined by people in the development of history. For example when it comes to the concept of new Asian style, even as an experienced designer, I cannot truly and profoundly understand the meaning of it. I'm not sure because the concept itself is defined by people or from a designer's inspiration when completing a project. What is Asia? China? Southeast Asia? Japanese? Korean? Is it clear? No one has a very clear answer. I guess the so-called design is to respect your service target, a way of expression to serve the inner self for the designer or one's soul journey with beautiful things.

I hope design can put aside the appearances of all forms and labels to present slight affects in life. Design appearance is to solve the relationships between things and human and thing while actually what we should solve is the relationship between humans. Finally we should solve the distance between hearts, the distance between human and the own heart.

Being in this industry for many years, I still love it, hope that those who are here can taste surprises and affects brought by designs and wish Design Vision International Publishing Co., Ltd (HK) to become better and better.

小隐于山 ，结庐而居，是中国古人热爱的一种生活形态。然而处于科技飞速发展的现代，能达到这种状态的人少之又少，更多的是于都市一隅选取喜爱的住宅而居，或者建设某一用于商业目的的场所，这就让室内设计在发展的过程中被不断的加重它的重要性。

作为一个70后，有幸目睹了这个时代的大规模建设。房地产的飞速发展，城镇化进程的日益提速，人们生活质量的显著提高，对室内设计的需求越来越彰显。设计行业的蓬勃发展所带给整个社会进步和生活质量的提升也是起到了至关重要的意义，室内设计在整个人类社会生存空间当中所存在的意义已是非凡。无建筑不设计，林立街道两侧的豪华五星酒店，霓虹闪烁的高级商场，格调优雅的主题餐厅，还有隐于市的各大豪宅，都隐秘地代表着现代人对生活的高品质追求和向往。

室内设计在中国的几十年，经历过无数的风雨，逐渐发展与成熟，形成了无数的世人所定义设计风格。然而越是趋于成熟，越是发现自身的不成熟。所有所谓的风格，其实都是在历史的发展过程当中被人们不断的定义。比如说新亚洲风，拿出这样一个概念，作为一个设计界的老江湖，是否真正的深刻理解他的含义，我不肯定，因为本身这个概念就是人为界定的，或是某个设计师在对某一设计完成时的一时的灵感迸发。何为亚洲？是中？是东南亚？是日？是韩？是否说的清楚？没有人会有一个很明确的答案。我想所谓的设计便是在尊重你的服务对象，或者是对设计者本人来说就是服务于内心真我的一种表达方式，也是自己心灵对于美好事物的一次旅程。

希望设计能抛开一切形式和标签的表象，呈现生活中细微的感动。设计表象是解决物和物以及人和物之间的关系，而实际我们应该解决的是人和人之间的关系，最后解决的是心灵与心灵之间的距离，人和自己心灵之间的距离。

混迹江湖无数载，仍然热爱着这个江湖，愿在设计行业的朋友都能够在工作中体会设计所带给自己的惊喜和感动，也祝愿香港视界国际出版社越办越好。

目录

C O N T E N T S

欧式风格 EUROPEAN STYLE

简欧风格 SIMPLE EUROPEAN STYLE

新中式风格 NEO-CHINESE STYLE

现代风格 MODERN STYLE

美式风格 AMERICAN STYLE

EUR
STY

OPEAN

LE

欧式风格

TOP MANSION CREATES REMARKABLENESS

顶级豪宅 缔造非凡

Design Concept | 设计理念

Walking near the house, a heavy European household sentiment attracts people deeply. The interior furnishings perfectly reflect the European noble palace style, giving people an illusion of seemingly being in a medieval mansion.

The furnishings in the mansion constantly strive for excellence. The luxurious ceilings embedded with gold foils, hollowed-out handrails, wall paintings, sculptures and fireplaces, all these artworks with sense of age make the entire space emit heavy artistic flavors and elegant charms.

走近这座房子，一种浓浓的欧式家居情调将人深深吸引。屋内的内饰完美体现了欧洲贵族宫廷风范，有宛若置身中世纪豪宅的错觉。

豪宅内的装饰可谓是精益求精。镶嵌金箔的奢华吊顶、镂空扶杆、壁画、雕塑、壁炉，有了这些颇具年代感的艺术品，整个空间散发出浓浓的艺术气息和典雅风韵。

项目名称：世茂天城
设计公司：福建品川装饰设计工程有限公司
设 计 师：周华美
项目地点：福建福州
项目面积：1000㎡
主要材料：壁纸、实木、挂画等

The living room uses large area of solid wood wall and beige carpet; the perfect collocation of two colors harmoniously integrates the entire space from the perspective of vision. The symmetric sofas on two sides are full of luxurious sense no matter on color or texture; the gold crystal droplight furnishings also become the excellent punchline; the collocation of these two adds beauty to each other. Dark green dining chairs add a retro color and exotic mysterious feeling for the dining room. The house is as exquisite as a treasure box and gives people unexpected surprise at any time.

客厅使用大面积实木墙面与米色系地毯，两个色系的完美搭配，将整个空间从视觉上十分和谐地融合起来。两侧对称的沙发摆设，无论从颜色还是质感上都充满华丽感，而悬挂上方的金黄色水晶灯饰也成了绝妙的点睛之笔，在搭配上更相得益彰。墨绿色的餐椅为内设餐厅增添了一抹复古色彩和异域神秘情调。房子就像一个精美的宝盒，随时给人预料之外的惊喜。

休闲区
客厅
书房
客卫
客房
书房
过道
更衣室
湿蒸房
公卫
玄关
衣帽间
干蒸房
书房
厨房
储藏间
棋牌室
电梯间
主卫
男孩房

浴缸
主卫
主卧
更衣间
书房
走道
过道
衣帽间
书房
电梯间
次卫
女孩房
健身区

In the master bedroom, the designer chooses solid wood wall; the entire space is in warm brown tone, in addition with bright white ceiling and classic purple soft coverage, presenting a unique flavor. Floral painting frame on the wall and retro furniture, all details manifest the owner's remarkable taste.

Compared with the master bedroom, the subaltern room gives people a deep impression of northern European style. The brush of forest green gives people pure and comfortable feeling. Bringing retro furniture into modern northern European space is a popular and fresh mixed style. Simple and funny decorative paintings bring unexpected happiness and vigor for the home.

在主卧室中，设计师选择实木的墙面，整体空间呈现温馨的棕色调，加上明亮的白色天花板，点缀经典的紫色系软包，别有一番味道。墙面上的碎花画框，复古的家具，无处不在的细节彰显主人的不凡品味。

相较于主卧，次卧的格调给人一种北欧风格的深刻印象。增添的一抹森林绿，给人纯净和舒适的感觉。将复古家具融入到现代北欧空间中，是现在流行的清新混搭风格。简单有趣的装饰画，为居室带来出其不意的欢乐活力。

WESTERN STYLE, CHARMING SCENERY

西洋风情 仙然景逸

Design Concept | 设计理念

If when you look at a person, you look at his eyes, what about looking at a home? It may be its hallway. When opening the door, what you see is the painting with a strong visual impact collided by yellow and blue. It is inspired by the painting *Irises* from Dutch painter Vincent Van Gogh. Iris means "rainbow" in ancient Greek and is the national flower of France. According to the legend, when the first king Clovis of the Kingdom of France was baptized, god sent him Iris as a present. Iris is an ordinary plant, but painting endows it with wonderful image and color and eternal vitality, which is the praise to nature and yearning for a better life. Entering into the brilliant purple in the painting, it seems to be a group of butterflies dancing in the space, spreading and permeating into every space, which slowly wakes up space and gradually gets into a silence. With a chill, a fragrance and an exquisiteness, in the prosperous space, is there your happy shadow?

如果说，看一个人就看其眼睛神韵，那么看一个家呢？应该就是看其门厅吧！当打开门迎面而来的是黄蓝撞色所创造强烈视觉冲击的画作，它的灵感来自于荷兰画家梵高的一幅画作《鸢尾花》，鸢尾花在古希腊语中是“彩虹”之意，而鸢尾花更是法国的国花，因为相传法兰西王国第一个王朝的国王克洛维在受洗礼时，上帝送给他一件礼物，就是鸢尾。鸢尾花本是一种很平凡的植物，但画作赋予它精彩的形象与色彩以及永恒的生命力，这是对大自然的赞美，是对美好生活的向往。走入画面中灿烂的紫色，它似乎一群翩翩起舞的蝴蝶在空间中蔓延，渗入在空间的每一处，慢醒了空间，循序渐进的悄无声息。掬一寸星寒，捻一缕花香，寄一丝精细，这一室荼蘼盛开依旧繁华，是否有你欢喜的身影?

项目名称：贵阳中渝第一城A2样板房
设计公司：深圳市北牧空间设计有限公司
设 计 师：李靖云
项目地点：贵州贵阳
项目面积：269㎡
主要材料：饰面板、香槟色亮光漆、大理石、镜面、墙纸、木地板等

Cine

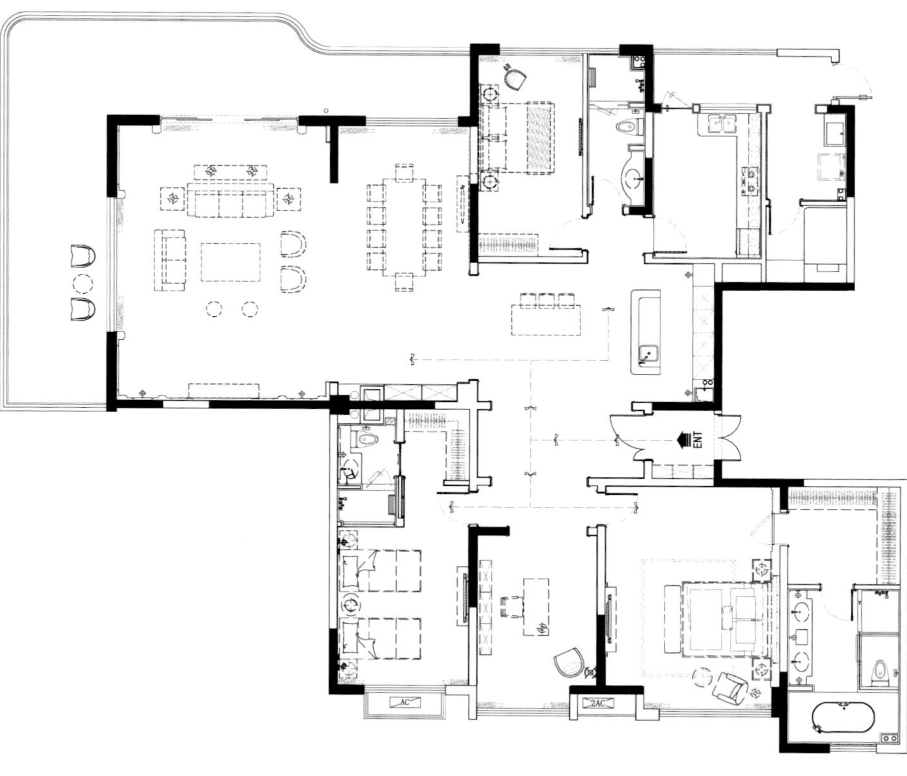

TIME FLIES, CLASSICS LASTS FOREVER

岁月流长 经典永存

Design Concept | 设计理念

The beautiful villa of this project quietly shines under the sun lights, resplendent and dazzling. Green grasses in the garden of the villa grow flourishingly; fresh flowers dance quietly with the wind. All is as beautiful as in the fairyland and cannot be touched.

The designer chooses elegant, noble, implicit, gorgeous, natural and harmonious neo-classical art style. Exquisite ceiling and large area stones promote texture of the entire space. Delicate jade beams and high fireplace match with noble and elegant European furniture, exquisite and ambitious. Heavy blue and leather texture convey the flavor of European style. European handmade sofa lines in every area are graceful with pretty colors. It pays attention to cloth color collocation and symmetry, manifesting noble identity of the residents with the beauty of aristocratic elegance and gorgeousness. The whole space emits elegant and fashionable flavors, making people linger on without any thought of leaving.

项目名称：长乐香江国际王公馆
设计公司：福建国广一叶建筑装饰设计工程有限公司
设 计 师：唐垄烽
项目地点：福建长乐
项目面积：600㎡
主要材料：大理石、墙板、仿古镜、壁纸等

本栋美丽的别墅静静地在阳光的照耀下，璀璨而夺目。别墅的花园中绿草茂盛地生长，鲜艳的花朵随微风轻轻地飘摇，一切都是那么美丽，仿佛是在童话中，让人不忍触碰。

设计师选用以优雅、高贵、含蓄、华丽、自然和谐为主的新古典艺术风格，精致的天花吊顶、大面积的石材提升整个空间的质感；精琢的玉石梁柱、大气的挑高壁炉等，配以高贵典雅的欧式造型家具，精致且富有气魄。浓烈的蓝调及皮质感更加传达出欧式风格的味道。而各区域里欧式手工沙发线条优美、颜色秀丽，注重布艺的配色及对称，彰显居者的高贵身份，十分具有贵族的高雅华丽之美。整个空间散发着典雅时尚的气息，让人流连忘返。

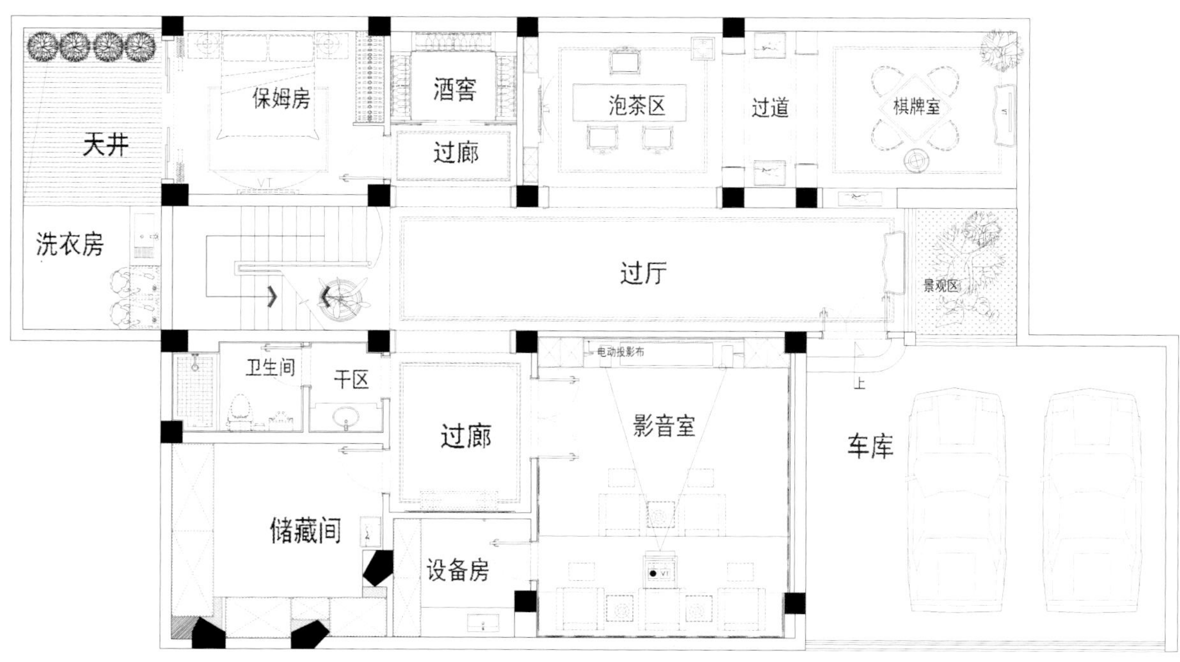

保姆房
酒窖
泡茶区
过道
棋牌室
天井
过廊
洗衣房
过厅
景观区
卫生间
干区
电动投影布
上
过廊
影音室
车库
储藏间
设备房

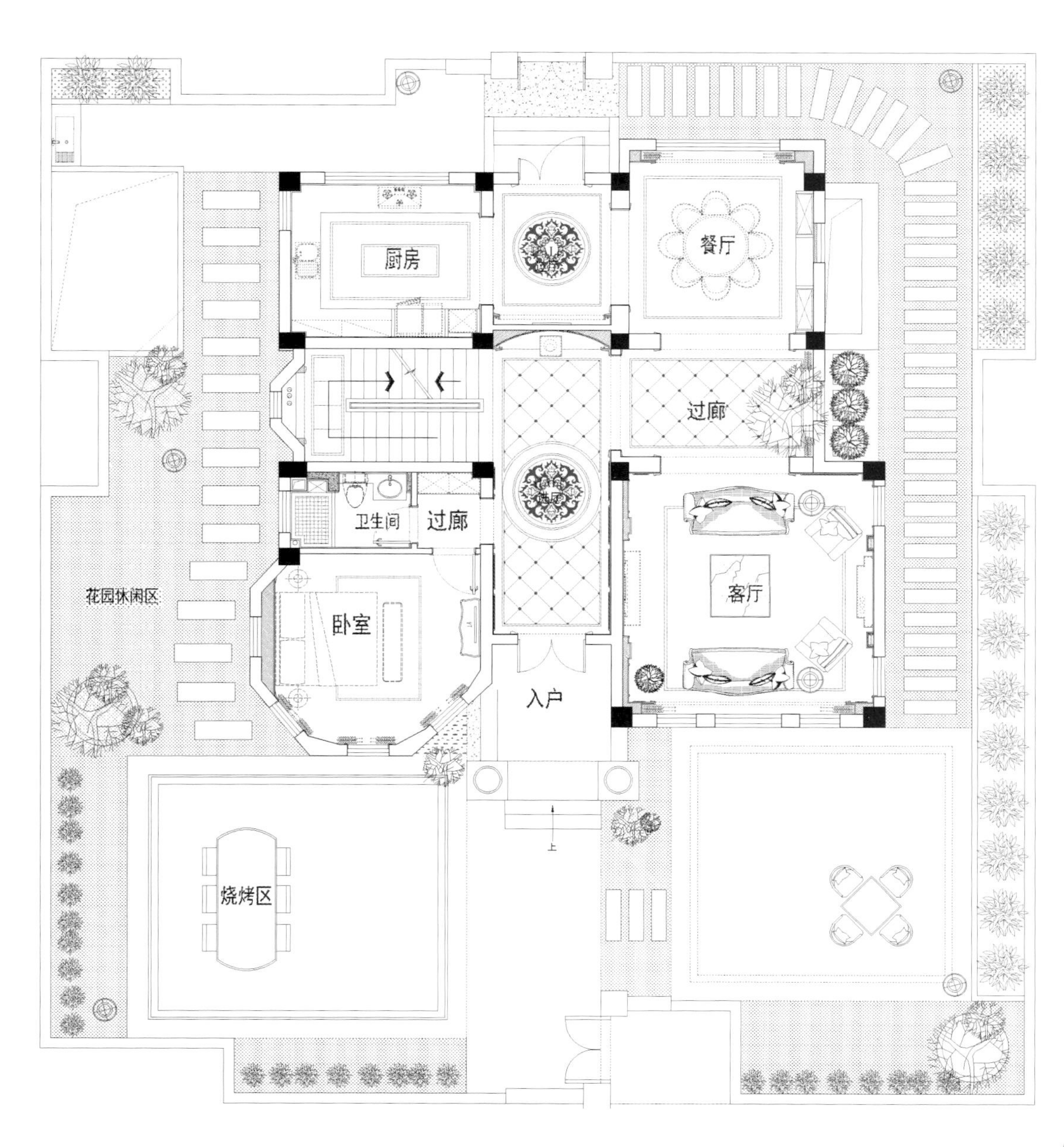
厨房
餐厅
过廊
卫生间
过廊
花园休闲区
卧室
客厅
入户
上
烧烤区

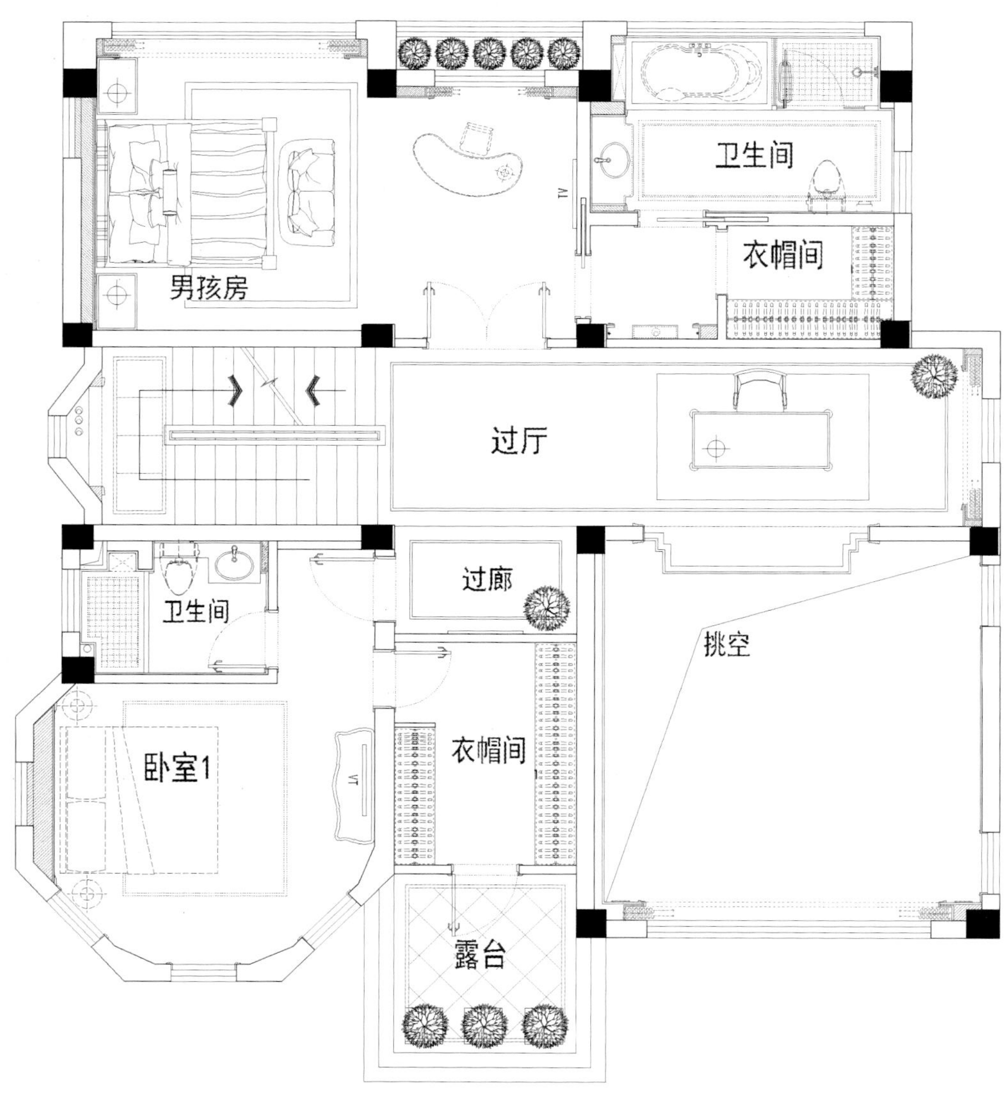
男孩房
卫生间
衣帽间
过厅
过廊
卫生间
挑空
卧室1
衣帽间
露台

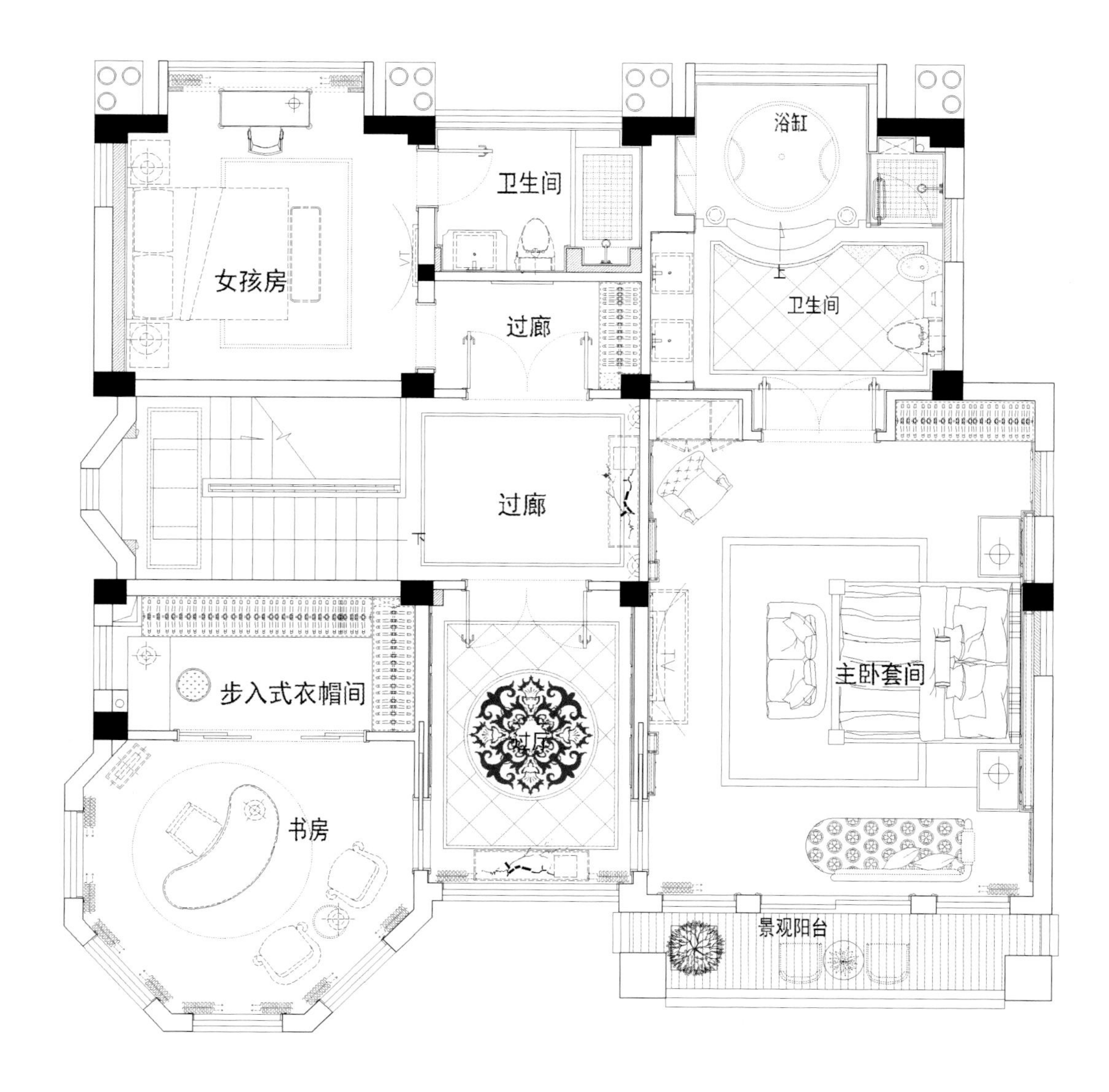
浴缸
卫生间
女孩房
卫生间
过廊
过廊
主卧套间
步入式衣帽间
过厅
书房
景观阳台

CHARMING STYLE, CHINESE AND WESTERN INHERITANCE

韵味格调 传承中西

Design Concept | 设计理念

The design style of this project is "the mix of Chinese and Western neo-classical style"; the whole space sets beige as the main tone; the collocation of wallpaper, partial ordered modeling and soft decoration creates an elegant, noble and luxurious living space. On details, classic European lines outline the modeling with rich layering; proper color contrast and exquisite detail nodes enrich the entire visual experience.

The soft decoration mainly chooses olive green neo-classical furniture, which injects a breath of fresh green into interior space, frees the residents from fatigue of work and makes their bodies and hearts back to life which is closest to nature. Curtain with vertical beauty, painting with European amorous feeling and classic fireplace create luxurious sense as the European court; the designer refines essences of Chinese style and integrates them into it; these two different design styles are deduced perfectly.

本案设计风格为“中西混搭新古典风格”，整个空间色调以米黄色为主色调，通过了墙纸、局部主副造型和软装的搭配，营造一个典雅、高贵、奢华的驻享空间。在设计细节上，以经典的欧式线条勾勒出富有层次感的造型轮廓，适当的色彩比对、精细的细节节点丰富了整体的视觉体验。

软装上主要采用了橄榄绿色调的新古典家具，一股清新的绿意注入到室内空间中，挣脱了工作的疲劳，把身心回归到最贴近自然的生活。带有垂直美感的窗帘、极有欧洲风情的挂画以及经典壁炉营造出欧洲宫廷的豪华感，设计师提炼了中式风格中的精髓融入其中，两种迥异的设计风格在此得到完美的诠释。

项目名称：碧桂园西苑别墅G295项目
设计公司：广州源创榀格环境艺术设计有限公司
设 计 师：周文胜
项目地点：广东佛山
项目面积：620m²
摄 影 师：戴胜念、梁锡坚
主要材料：扪皮、木饰面、木地板、仿石瓷砖、墙纸等

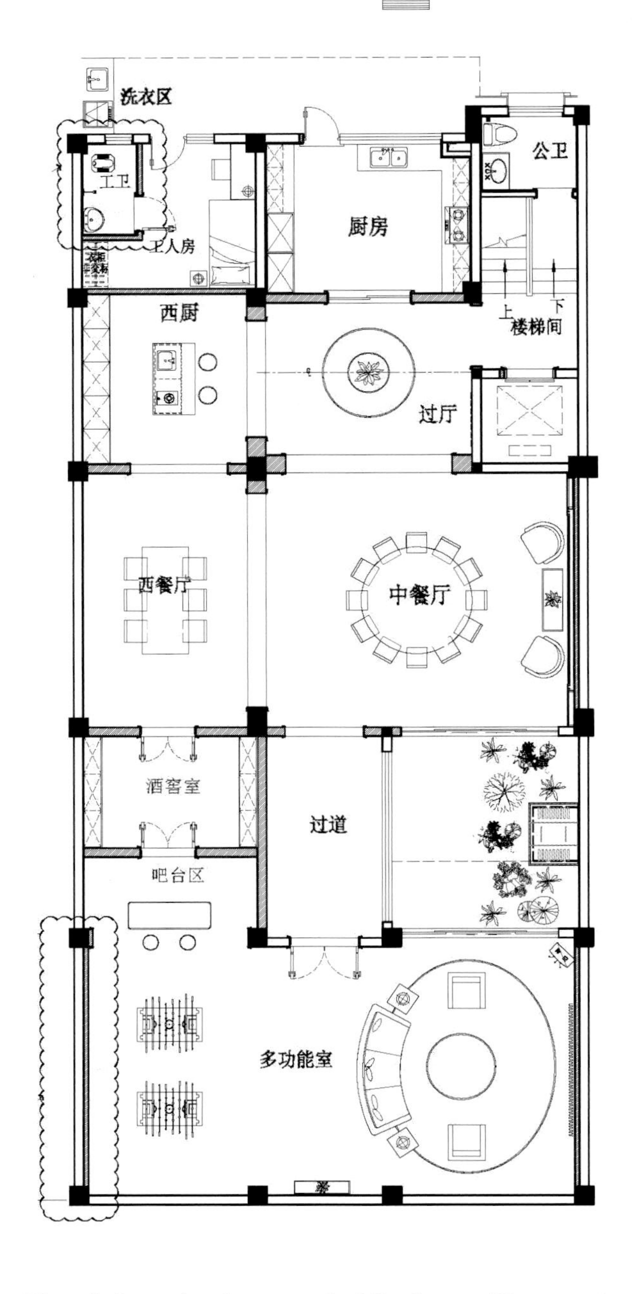

The whole project has a total of five layers. The negative layer is dining and entertainment spaces; the first floor is reception spaces; the other floors are bedrooms. The spaces are spacious and decent with stairs and elevators interior; the humanized designs promote the level of the whole space. The designer combines the customers' lifestyles and habits to create a unique private space for everyone; the shift of every scene seems to tell a different story and seems to contain unique meanings. Our work brings the clients a feeling of home and leads them to enjoy real comfortable life through professional design visions.

整个项目一共五层，负一层为用餐和娱乐空间，首层为会客厅空间，二三四层为卧室空间，空间宽敞大气，室内设置了楼梯和电梯，人性化的设计，提升了整个空间的档次。结合着客户的生活方式和习惯缔造出每一个属于他们独一无二的私人空间，每个场景的转移似是倾诉着不同的故事，又似乎蕴含着独有的意义。我们的作品带给客户一个家的感觉，同时用专业的设计视野带领他们享受真正的舒适生活。

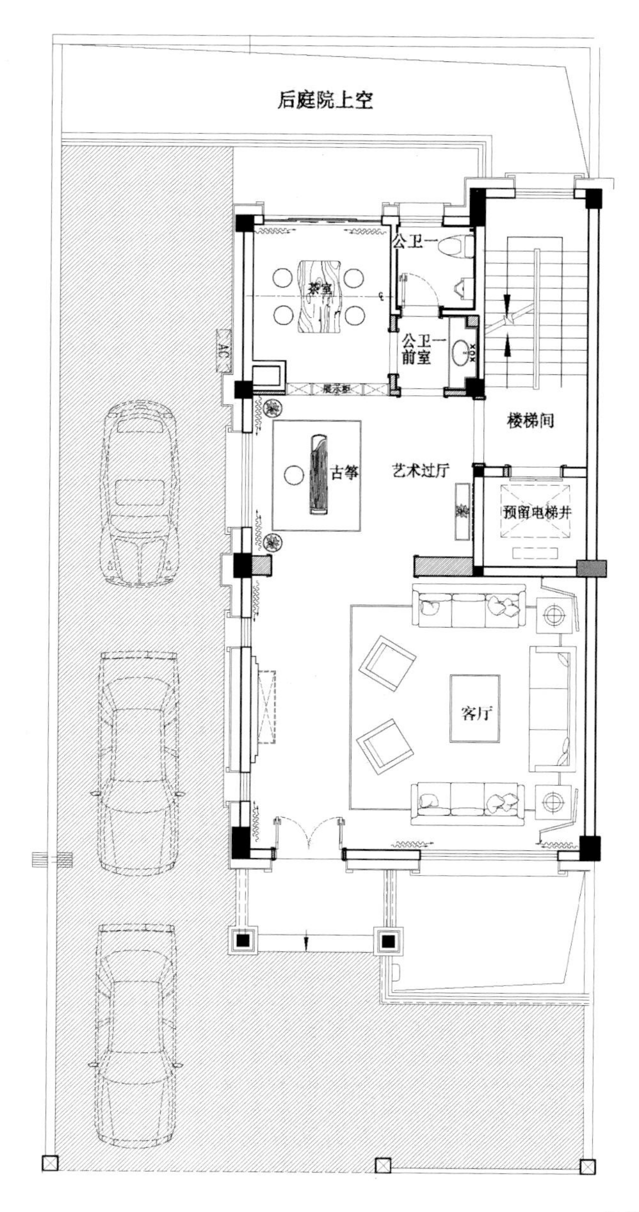
后庭院上空
公卫一
茶室
公卫一
前室
AC
展示柜
楼梯间
古筝
艺术过厅
预留电梯井
客厅

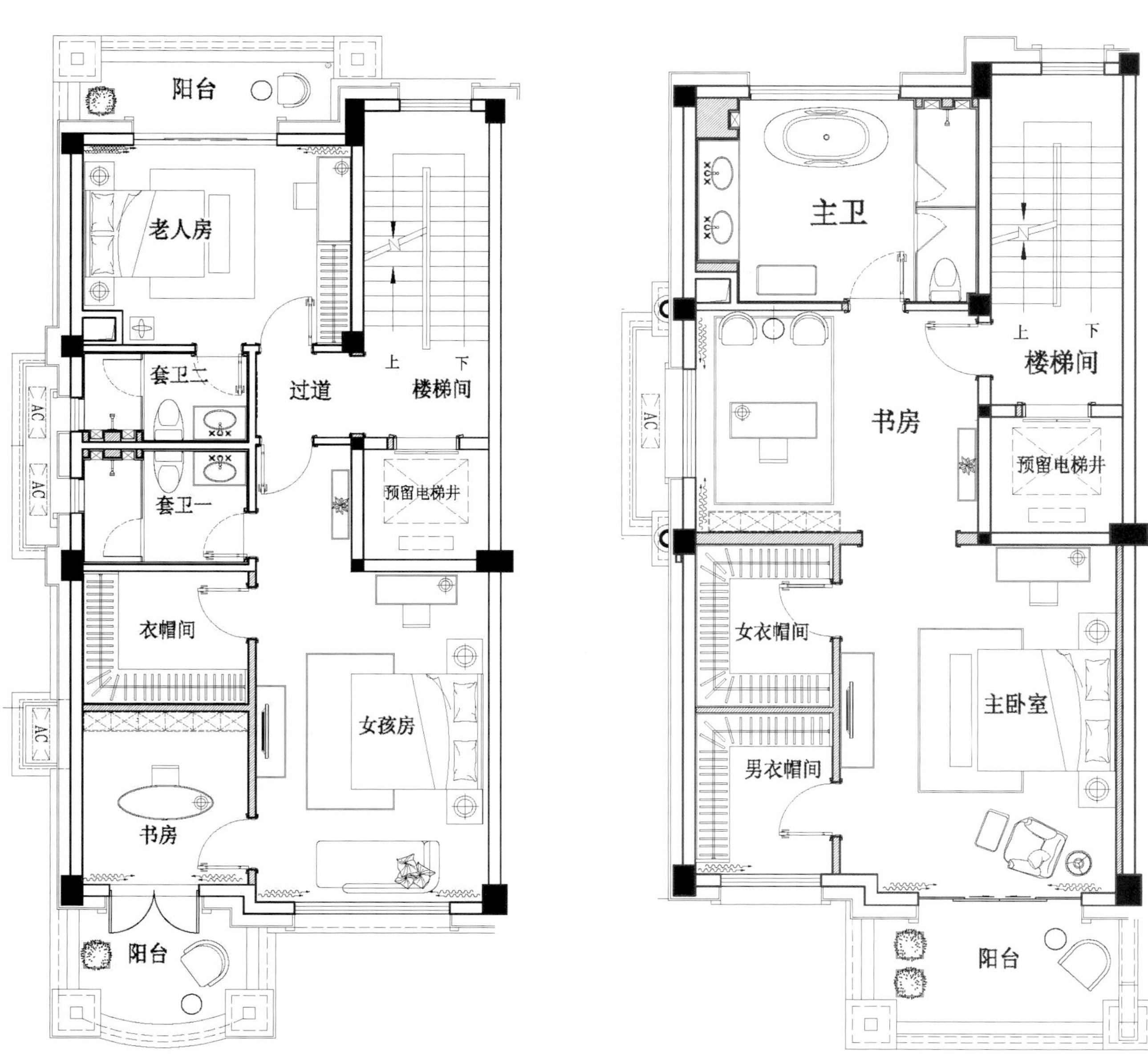
阳台
老人房
套卫二
过道
上
下
楼梯间
套卫一
预留电梯井
衣帽间
女孩房
书房
阳台
AC
AC
AC
主卫
上
下
楼梯间
书房
AC
预留电梯井
女衣帽间
主卧室
男衣帽间
阳台

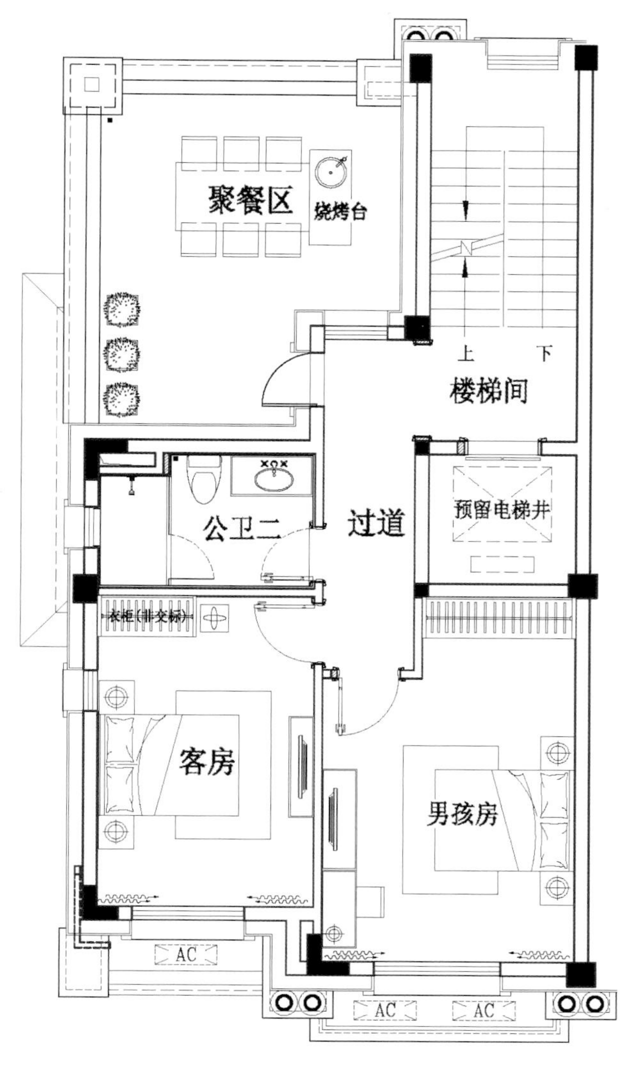
聚餐区
烧烤台
上
下
楼梯间
公卫二
过道
预留电梯井
衣柜(非交标)
客房
男孩房
AC
AC
AC

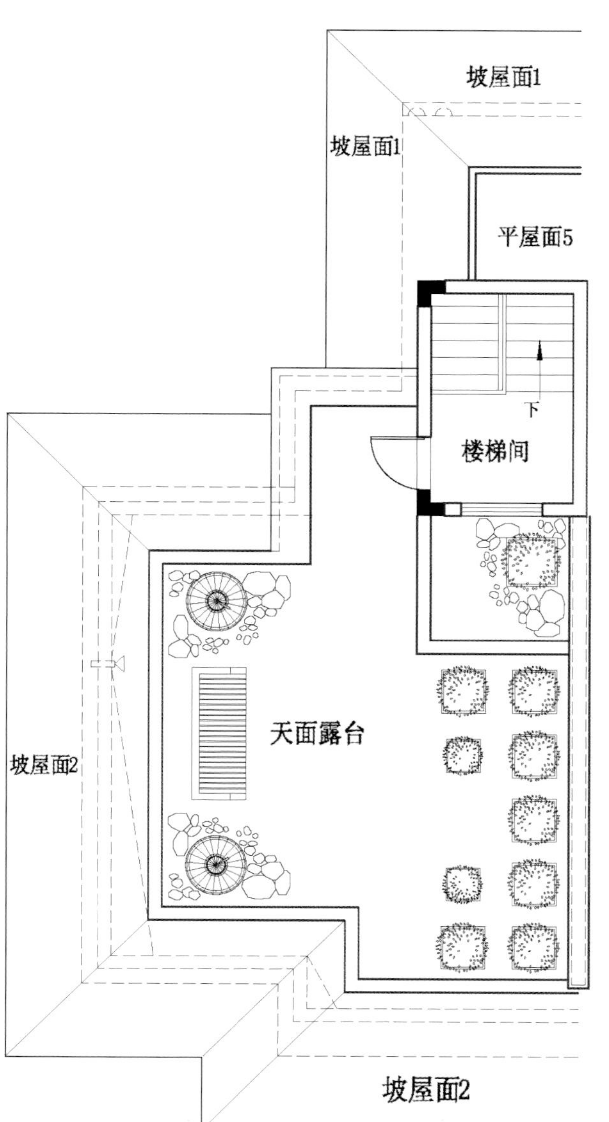
坡屋面1
坡屋面1
平屋面5
下
楼梯间
天面露台
坡屋面2
坡屋面2

NEW CHANT OF CLASSICISM

古典新咏

Design Concept | 设计理念

The architectural layout of this project stresses symmetric effect of axis and uses classical beauty of symmetric harmony to create magnificent and spacious manner, set up luxurious and comfortable living atmosphere and manifest noble and elegant temperament. The high living room is decent and spacious; graceful and luxurious droplight in the dining room is dazzling; the delicate and exquisite bedroom seems to weave a lively and beautiful dream for the owner. The style is luxurious, gorgeous and full of romance and movement. The architectural details use French colonnade, carving and lines; the manufacture crafts are delicate and exquisite. The natural arrangement with casual furnishings advocates beauty in conflict; all kinds of colors set off each other and create fresh and romantic feelings.

本案在建筑布局上突出轴线的对称效果，十分具有古典的对称和谐美，借以营造恢宏广阔的气势，布置出豪华舒适的居住环境，彰显贵族般高贵典雅的气质风格。挑高的客厅大气开阔，雍容奢雅的餐厅吊灯惹人注目，精致讲究的卧室仿佛为主人编织了一个明快美丽的梦想。整个风格豪华、富丽，充满浪漫和动感。在建筑的细节处理上，运用了法式廊柱、雕花、线条等，制作工艺极为精细考究。自然摆设中不经意点缀，崇尚冲突中产生美，以各种色彩的互相映衬创造清新浪漫的感觉。

项目名称：芙蓉别墅
设计公司：浙江省温州市天甫装饰设计有限公司
设 计 师：吴建
参与设计：吴达、高升
项目地点：浙江温州
项目面积：1500㎡
主要材料：大理石、布艺、墙纸、原木等

ELEGANT LIFE BEYOND LUXURY

奢华之外 高雅生活

Design Concept | 设计理念

The owner of this project prefers luxurious space and noble lifestyle and highlights elegant and connotative life temperament, so the designer focuses on presenting taste and luxury and reflecting unique visual effect to make the entire textures contrast strongly and lines soft and hard. A piece of "long flowing stream" manual painting of the living room, fountain spindrift top, jade belt carved column stair, 8-shaped hall, trumpet-shaped wealth-gathering entrance and gold ingot hall, "material" and "form" of every detail design are carefully thought. Warm, luxurious, noble, gorgeous yet not complicated, concise yet not simple, it not only meets residents' pursuits of life quality and nobility, but also manifests life taste. At the same time, Chinese people pay attention to good morals and symbols, and this work is the existence of the design thought and essence.

本案业主倾心奢华空间与贵族生活的方式，注重高雅与内涵的生活气质，因此设计师注重彰显品味、奢华以及体现独特的视觉效果，使整体质感对比强烈，线条更是柔硬并存。其中客厅的一副“细水长流”手工画、喷泉浪花顶、玉带盘龙柱楼梯、8字形如意厅以及喇叭状的聚财进口配以元宝入户门厅……在每一个细节设计的取“材”及造“形”上都是深思熟虑、精益求精。温馨、豪华、高贵、华丽而不繁琐、简约不失简单，不仅满足了居住者对生活品质与贵气的追求，也彰显出内质的生活品味。同时，中国人讲究好的寓意及象征，这套作品便是体现其设计的思想与精华所在。

项目名称：山水黔城府邸别墅
设计公司：广东星艺装饰集团贵州有限公司
设 计 师：罗山锐
项目地点：贵州贵阳
项目面积：600㎡
摄 影 师：誉邦影像工作室钟康学
主要材料：微晶石、实木护墙板、进口墙布、大理石、进口肌理漆、手工油画、艺术PU线、真皮软包等

BEAUTIFUL COLORS IN THE WORLD

浮世华光

Design Concept | 设计理念

The integral style of this project is European style. Walls of public spaces are covered with marble slabs and marble columns to build traditional European modeling and create exalted and magnificent space temperament. Top surfaces of conditioned spaces are made with arched or domed roofs, further promoting the sense of space and stereo. The floors use polished tiles to lighten the space, giving people open and extraordinary space feelings. The designer focuses on the choices of materials; expensive carpet and wide and exquisite furniture match with elaborately carved flower patterns, manifesting elegant and noble temperament of the space.

案例整体风格定位为欧式风格，公共空间的墙面用大理石板及大理石柱塑造出传统欧式造型，营造尊崇大气的空间气质；顶面在有条件的空间都做了拱形或穹形顶，进一步提升空间感和立体感；地面用抛光的地砖提亮空间，给人以开放、非凡的空间感受。设计师着重于材质用料的选择，价格不菲的地毯，宽大精美的家具，配以精雕细琢的花纹，空间高雅尊贵的气质展露无遗。

项目名称：西郊庄园
设计公司：微风设计
设 计 师：吴钒
项目面积：1500㎡
主要材料：大理石、木地板、地砖、壁纸等

On space division, the first floor is reception spaces including hall, living room, dining room, kitchen and VIP room; the negative floor is recreational spaces including living room, recreational room and video room; the second floor is children's room; the third floor is master bedroom with study and reception room. The dynamic and static areas are divided obviously and well-equipped, which not only fully meets the owner's using functions, but also reflects the noble and elegant living feelings.

空间划分上，平街的一楼作为接待空间，设置门厅、客厅、餐厅、厨房、贵宾厅等；负一楼作为娱乐空间，设置娱乐室、影音室等；二楼作为小孩的房间；三楼作为主人房间，配备书房、接待室等。动静分区明显，配置齐全，不仅充分满足了屋主的使用功能，而且体现出尊贵又不失优雅的居家情调。

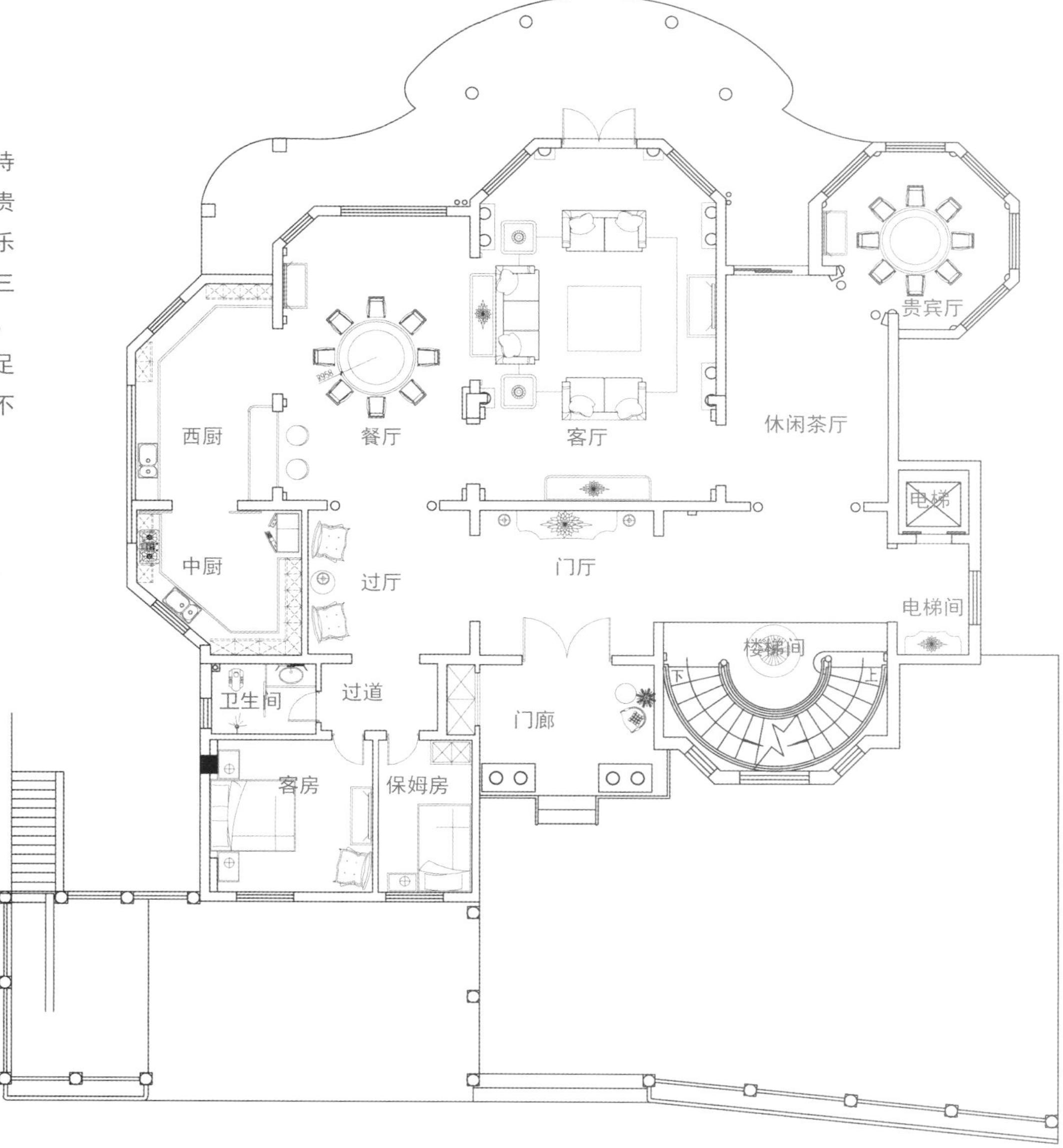

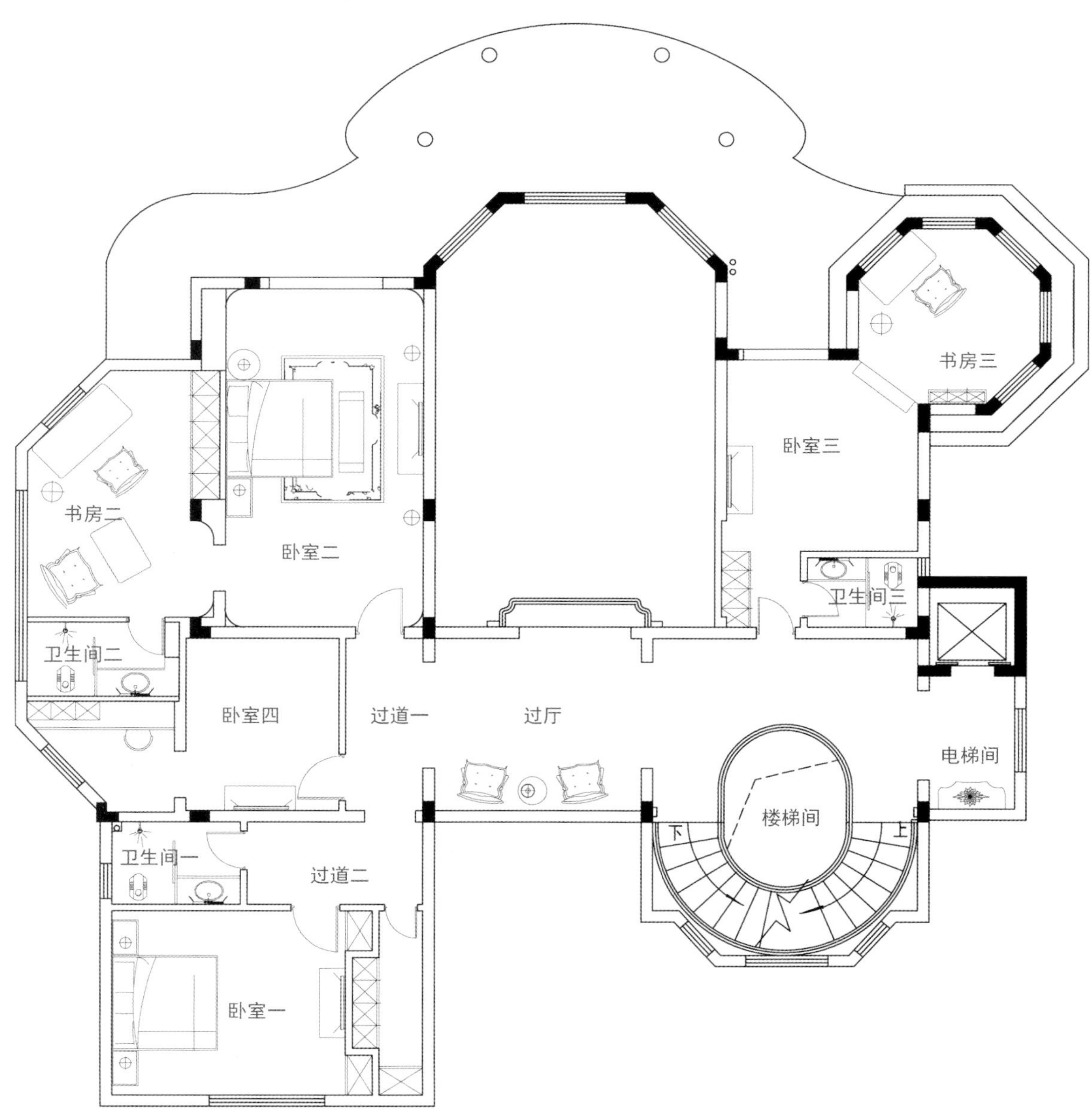
书房三
卧室三
书房二
卧室二
卫生间三
卫生间二
卧室四
过道一
过厅
电梯间
楼梯间
下
上
卫生间一
过道二
卧室一

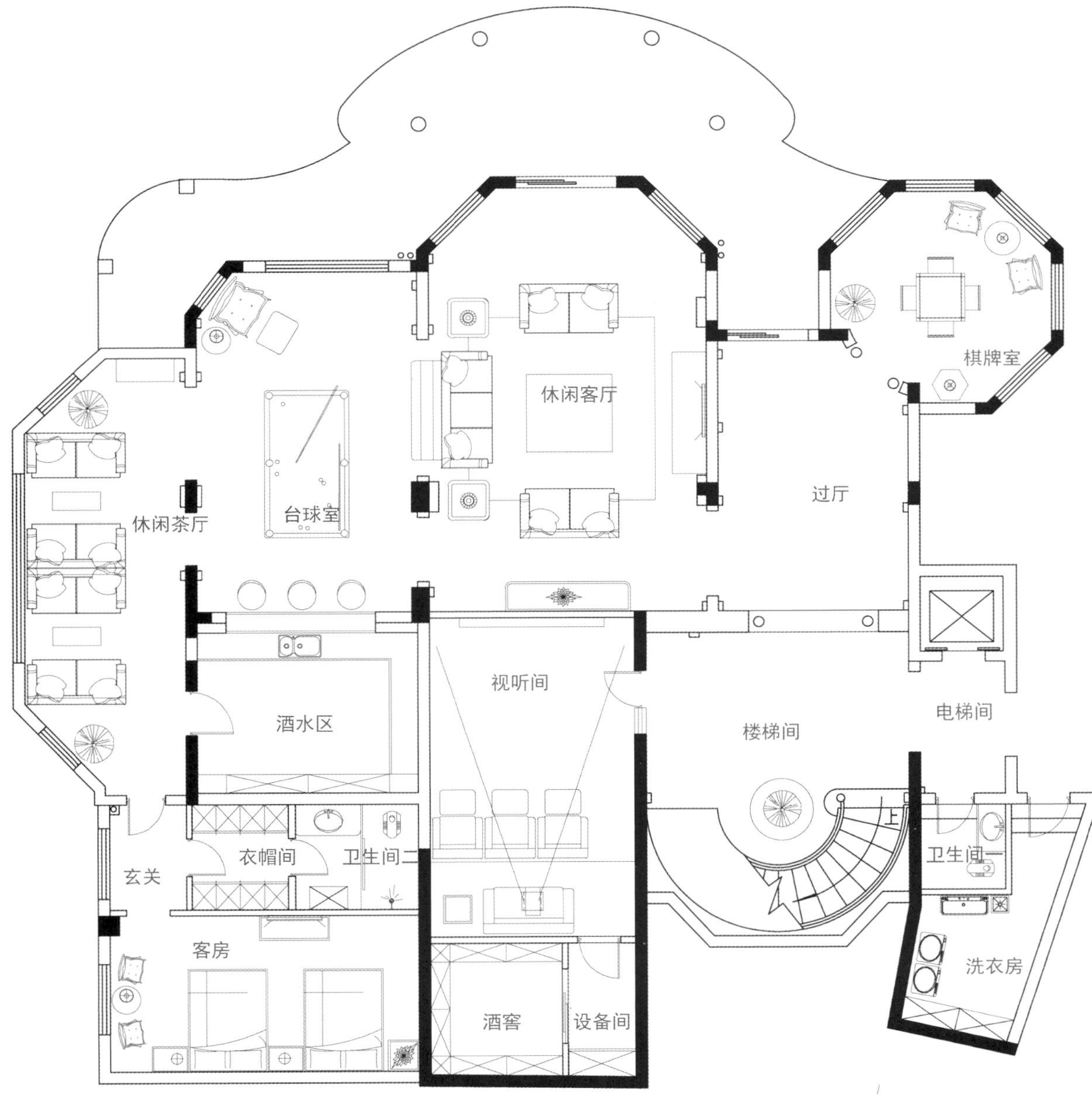

棋牌室
休闲客厅
台球室
休闲茶厅
过厅
视听间
酒水区
电梯间
楼梯间
上
衣帽间
卫生间二
卫生间一
玄关
客房
洗衣房
酒窖
设备间

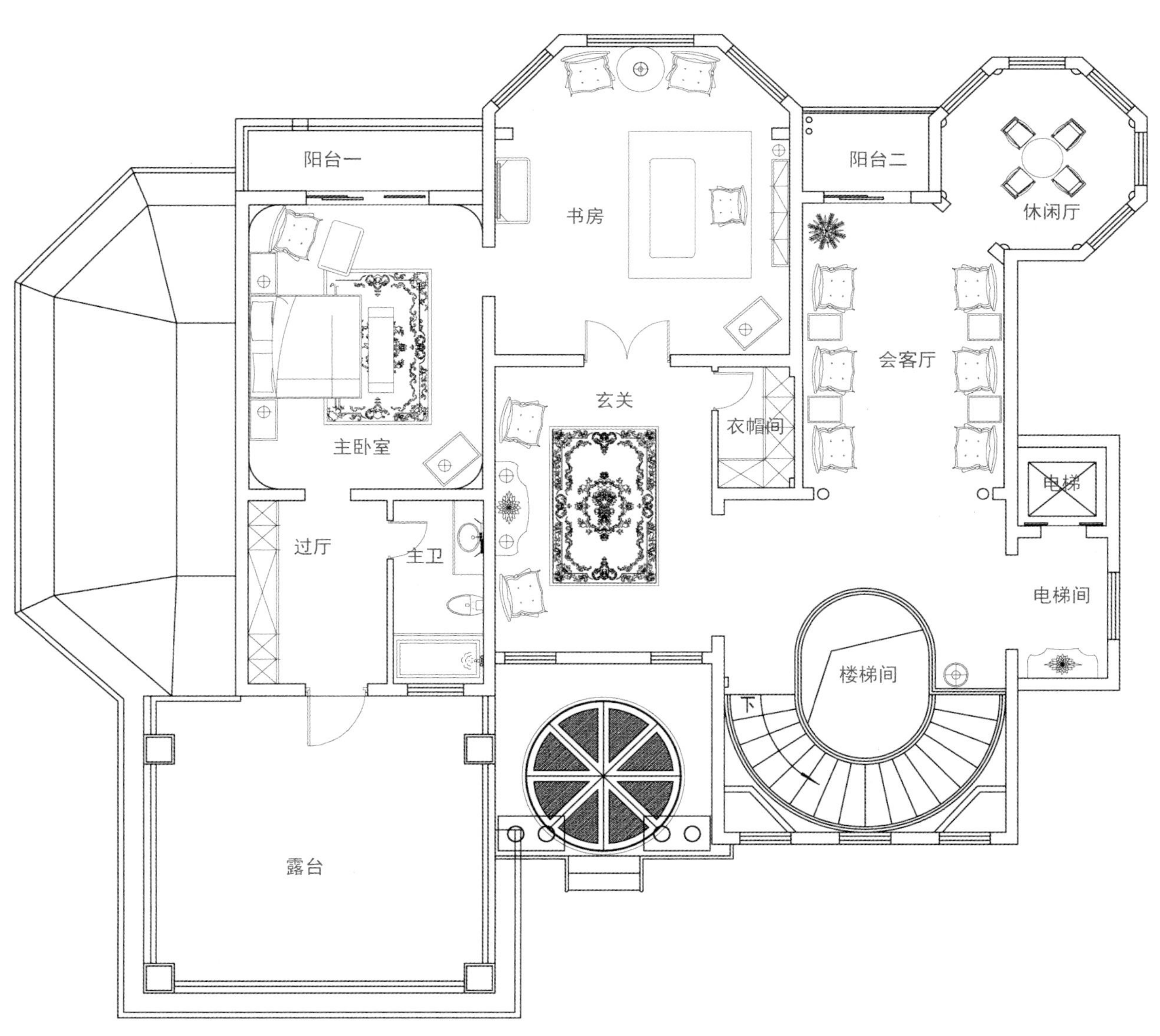
阳台一
书房
阳台二
休闲厅
会客厅
玄关
衣帽间
主卧室
电梯
过厅
主卫
电梯间
楼梯间
下
露台

SIMPL
EURO

简欧风格

PEAN STYLE

ANCIENT AND MODERN ENCOUNTER, ROMANTIC AND ELEGANT

古今相遇 浪漫优雅

Design Concept | 设计理念

Haihang Mansion is a luxurious mansion of Haihang real estate, occupies the leading place of Daying Mount international tourism island CBD, watches the future of CBD metropolitan prosperity, is a core place with functions including government affairs, commerce, transportation and life and provides livable and beautiful life bay for middle class elites. The designers abandon complicated and heavy classical European style to simplify and create and combine with urban elements to bring the younger generation vigorous modern simple European style. Black gold foil plum flower pattern side cabinet is as retro as the valuable possession of the European aristocrat in the last century which crosses the ocean to here, skillfully collocating with a piece of modern painting, new and old, modern and retro, colliding a modern European sparkle. Modern European style appears in front of us in a new gesture and depicts the elegance of European culture, modern and fashionable.

项目名称：海口·海航豪庭北苑三区C户型
设计公司：SCD（香港）郑树芬设计事务所
主案设计师：郑树芬、徐圣凯
软装设计师：杜恒、陈洁纯
项目地点：海南海口
项目面积：293㎡

海航豪庭是海航地产开发的实力豪宅，独占大英山国际旅游岛CBD龙头之位，守望CBD大都会繁华未来，是集政务、商业、交通、生活等多功能为一体的核心领地，为中产精英阶级构建宜居至美生活湾区。摒弃繁复厚重的古典欧式，进行简化创造，结合都市元素为年轻一代带来充满活力的现代简欧设计。黑色金箔梅花纹边柜复古的仿佛上世纪欧洲贵族的名贵家当漂洋过海而来，巧妙搭配一幅现代画作，新与旧，现代与复古，碰撞出现代欧式的火花。现代欧式风格以一种崭新的面貌呈现在我们面前，描绘出欧洲文化的优雅，又不失现代的时尚性。

ICE AND FIRE

冰与火

Design Concept | 设计理念

In this space, we can not only find rational and restrained noble temperament, but also see luxurious and hedonic colors. Through the contrast between furnishings and space, the passionate art breaks the rational tranquility and harmony and presents a heavy romantic feeling. The designer seems to use the technique of expression of Baroque art to create an artistic living space. From the point of design elements, light color background wall, a little exaggerated furniture, metal and rock furnishing materials, characteristic decorative elements and dramatic design works release the owner's inner romance in a soft and elegant way and present a more dramatic and profound sense of space in the visual contradiction.

在这个空间里，我们既能发现理性内敛的贵族气息，又可以看到豪华与享乐主义的色彩。通过陈设与空间的对比，以激情的艺术，打破了理性的宁静和谐，呈现浓郁的浪漫主义色彩，设计师似乎在用巴洛克艺术的表现手法，塑造一个艺术化的生活空间。从设计元素来看，浅色的护墙背景、略带夸张的家具、金属与岩石肌理的配饰材质、个性的装饰元素以及富有戏剧性的设计作品，以一种柔和、高雅的方式释放着主人内心的浪漫，并在视觉矛盾中呈现更具戏剧化的、更为深刻的空间感官。

设计公司：金元门设计公司
设 计 师：葛晓彪
项目地点：浙江宁波
项目面积：400㎡
摄 影 师：刘鹰
主要材料：装饰画、护墙板、雕塑等

HUMANISTIC CONNOTATION, ELEGANT BLACK-BLUE DEVELOPMENT

人文底蕴 典雅黛蓝

Design Concept | 设计理念

"Charming with light make-up or dressing in white, the mountains seem green with river clear as the blue sky". The whole design ranges from black-blue development to azure glaze popular in scholars; the western architectural space is integrated with the eastern noble feelings; the eastern aesthetic standard and the western spatial charm, combining with naturalism, have created a grand, elegant and affectionate space. The restrained and peaceful mind is intellectual with high quality, elegant yet not artificial.

Following the layout of mansion from front courtyard to backyard garden for scholars, the space is designed with four floors, and the progressive structure conforms to traditional dynamic living habits and psychological care, and it borrows the eastern philosophical thought, wisdom and skills of craftsmen to explore the optimal residential space to meet the demand of modern people.

“淡妆素裹总相宜，晴山如黛水如蓝”。本案整体以文人喜欢的黛蓝退韵到天青釉，在西方的建筑空间内融入东方的贵族情怀，东方的审美观点和西方的空间韵味结合自然主义的美学精神，塑造一个大气典雅而富有情怀空间。内敛平和的心境，知性不失品质，文雅不造作。

空间共计四层，规划遵循文人府邸（前庭--后花院）的布局依据，从递进布局上尊重传统的居住动线习惯及行为心理关照，运用东方的哲学思想、智慧，以匠人之心探究最适宜当代人需求的居住空间。

项目名称：康桥林溪湾603别墅样板间示范单位
设计公司：筑详设计机构
主创设计师：刘丽、张岩
项目地点：河南郑州
项目面积：620㎡
摄 影 师：耿旭姗
主要材料：云朵拉灰石材、大花白石材、香槟金金属板、实木木地板、米灰色定制墙板、刺绣玻璃屏风、香槟金箔油画等

In the first floor, traditional ceremonial layout starts with nobleness and glory, and the cares of functional units break through the sense of formalism, and it pays more attention to the dynamic living habits and needs of behavior psychology. An balanced visual effect is achieved within the frame of ceremony; the designs of scenes vary with walking and interaction and spatial shift in halls; rooms and atriums focus on the support of ceremonies and make the space full of oriental impressionistic feelings and styles of mansions; the atriums going from up to down and the free combination of west style symbols enable off-duty owners to achieve psychological transmission and relaxation; the living room with open to below is designed with main color to show the nobleness of the mansion; the interactive designs between home studio, living room, the elderly' s suite, dinning hall and exterior courtyard originate from the irresistible needs of the people to the nature.

一楼传统的仪式感布局开启府邸的尊贵与荣耀，功能节点的关照打破形式感的表象，更加注重居住动线习惯及行为心理的需求，在仪式感的框架内追求均衡的视觉效果；堂、室及中庭的移步换景及对景的设计及转换空间注重仪式感的烘托，使空间更具东方写意情怀和大宅风范；中庭自上而下的贯穿，西式符号的自由组合，让在归家后进行心理上的过渡和放松；挑空的起居室配以主体色调的着重刻画“府邸”的尊贵感；书房、起居室、老人房及餐厅与外庭院的互动设计来源于人对自然不可抗拒的诉求。

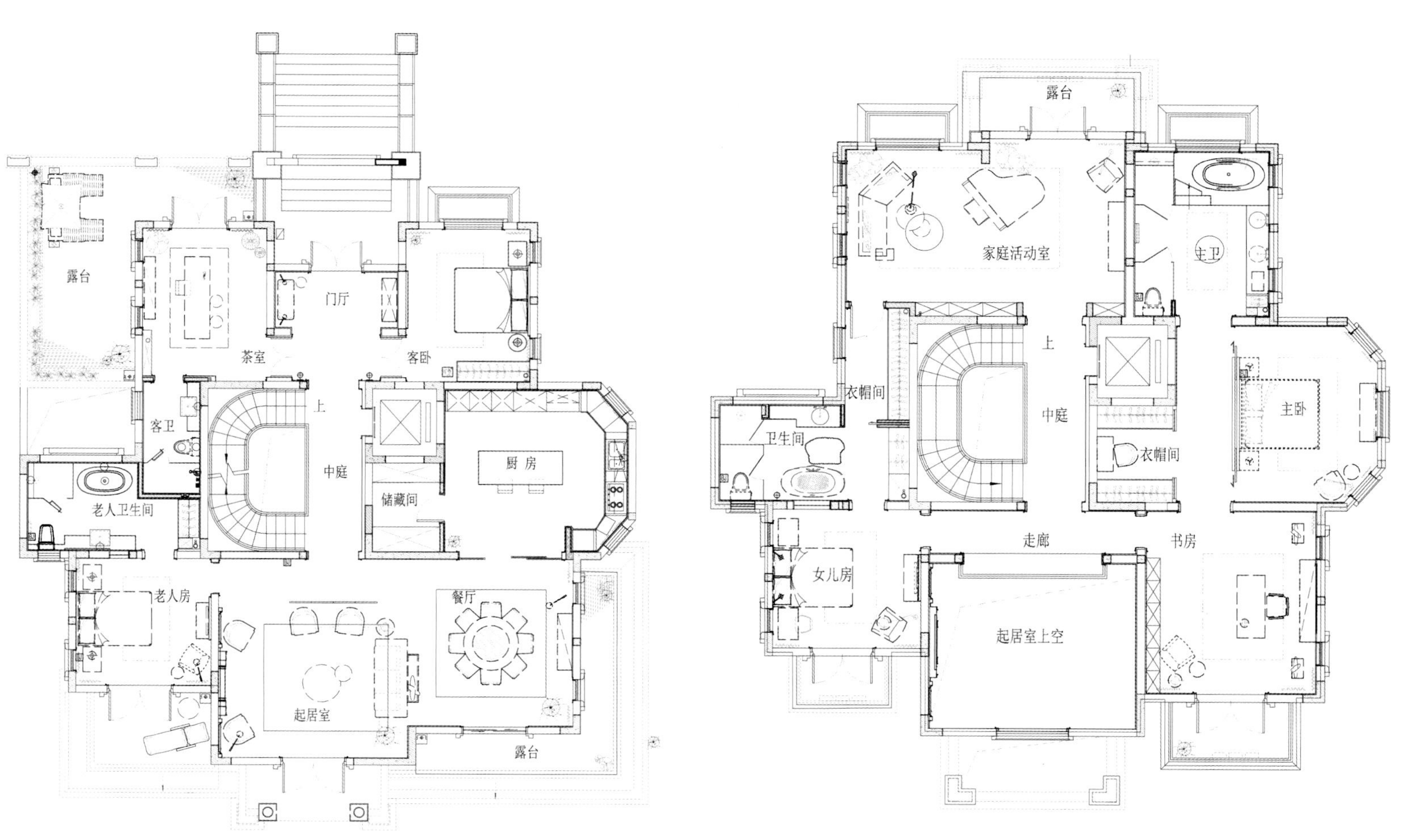
露台
茶室
门厅
客卧
客卫
上
中庭
厨房
老人卫生间
储藏间
老人房
餐厅
起居室
露台
露台
家庭活动室
主卫
上
衣帽间
中庭
主卧
卫生间
衣帽间
走廊
书房
女儿房
起居室上空

The family room is shared by all family members, where female friends may be invited, family affections be thickened and all can enjoy the family happiness; the master bedroom faces a window through which the scenes outside are visible, besides, it is equipped with a private dressing room and wide toilet, quiet and comfortable, just like a holiday spending.

家庭活动室是家庭和谐的共享场所，这里可以邀约闺蜜、浓化亲情、享受天伦之乐；主卧室面窗而居，将居住者的视线拉向室外景观，配以专属的私人更衣间及宽敞舒适的主卫，恬静舒适、有种度假惬意的生活情趣。

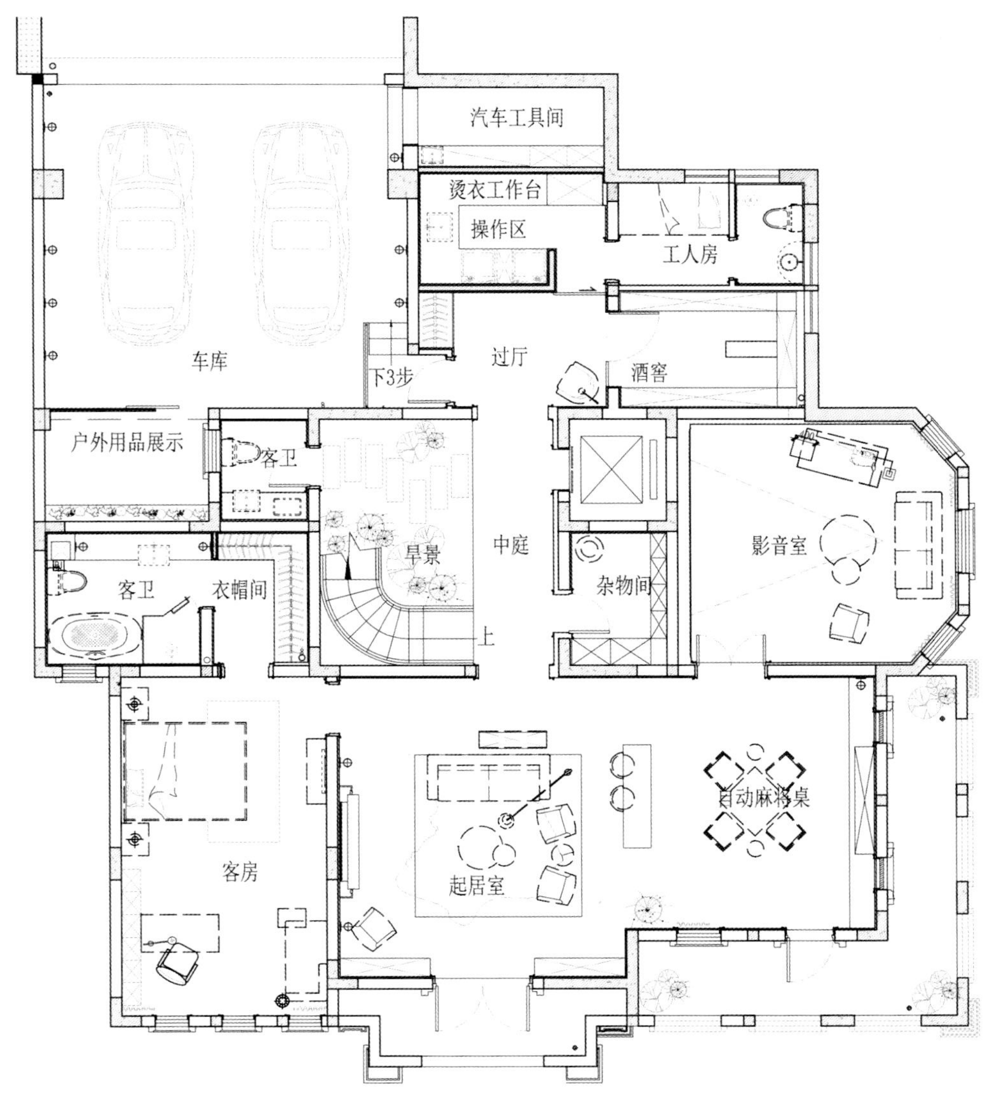

汽车工具间
烫衣工作台
操作区
工人房
车库
过厅
下3步
酒窖
户外用品展示
客卫
旱景
中庭
影音室
客卫
衣帽间
杂物间
上
自动麻将桌
客房
起居室

Mental nourishment for material sublimation is secrete and private; it regards the sacrifice of host after material enjoyment, which is an embodiment of status and grade; re-arrangement of locker rooms for mother and daughter and rational care design originate from the needs of the owner in actual living. Arrangement of dynamic living lines including parallel design of service passage and host's passage avoids awkward meeting in daily life; workers' houses are set at the entrance of garage to make it easy for out-going and internal service, and simple reception space is provided to care for the life of workers; private spaces for the host such as horsemanship, red wine, cigar, reading, walking inside the garden, feeling expression, the bigger space for hobbies and so on, manifest the commercial status of the host.

物质升华的精神食粮，私密专属；更注重男主人对物质享乐后的祭奠，身份和品位的体现；母亲衣帽间及女儿衣帽间的重复配置，合理的功能关怀源于居住者实际生活的需求落地；注重生活动线的划分，服务通道和主人通道的平行处理，避免尴尬的生活碰撞；工人房设定在车库入口处，方便外出及内部空间的服务，在车库还准备有简易的接待空间，关照工人的生活诉求；主人的私属空间：马术、红酒、雪茄、阅读、游园、性情抒发，更大的爱好的释放空间，彰显主人的商务身份。

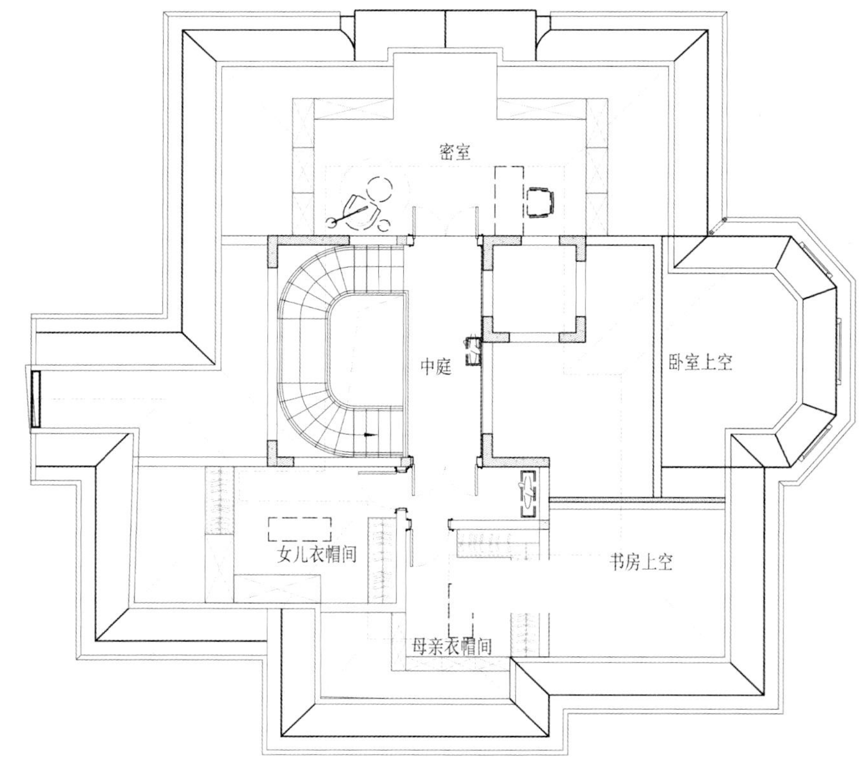

CAESIOUS PLACE IS PLEASANT TO LIVE

青灰染处怡人居

Design Concept | 设计理念

Facing with busy work and fierce competition pressure, modern people are yearning for a peaceful, tranquil, bright, capacious and comfortable home to help them eliminate the tired work and forget the noisy city, encountering with sunshine in the afternoon and taking off the splendid clothes. The European noble elegance is as before and gives a clear back shadow in the space; light color floor and blundering mood deposit together in the sunshine like gauze.

现代都市人面临着忙碌的工作，激烈的竞争压力，更向往一个安静、祥和、明朗宽敞舒适的家，帮助我们消除工作的疲惫，忘却都市的喧闹。午后与阳光于此邂逅，褪去富丽堂皇的衣裳。欧陆的贵族风华依旧，空间透出清澈的背影，浅色的地板与浮躁的心情一起沉淀在如薄纱般的阳光里。

项目名称：绿地·乔治庄园别墅
设计公司：郑州青草地装饰设计有限公司
设 计 师：直涵明、李君岩
项目地点：安徽合肥
项目面积：600m²
摄 影 师：耿旭珊
主要材料：石材、成品板、乳胶漆、壁纸等

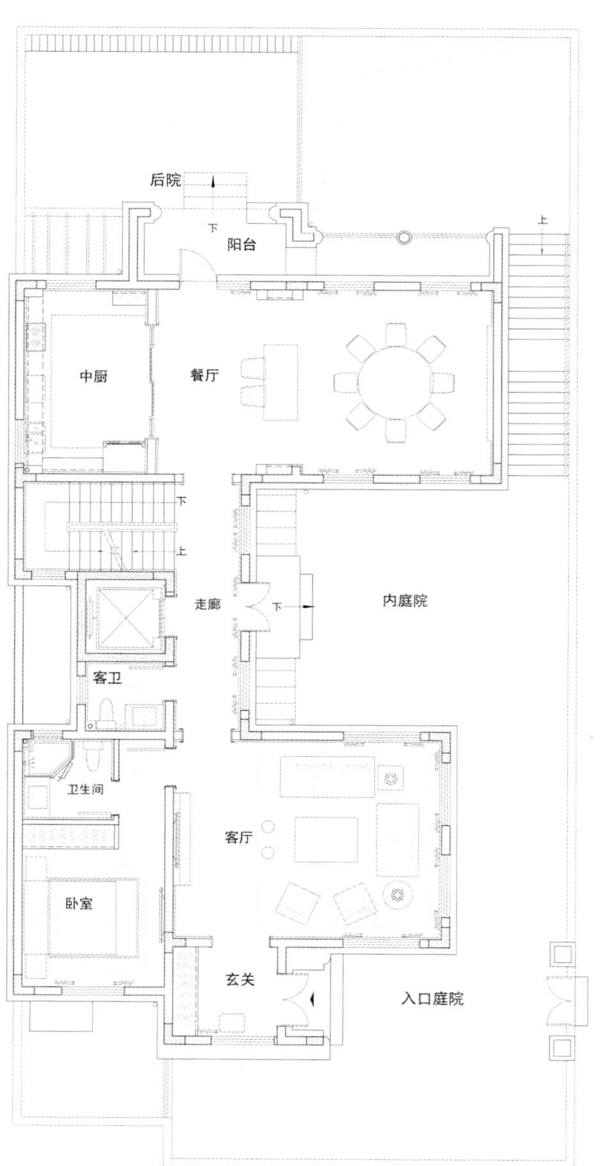
后院
下
阳台
上
中厨
餐厅
下
上
走廊
下
内庭院
客卫
卫生间
客厅
卧室
玄关
入口庭院

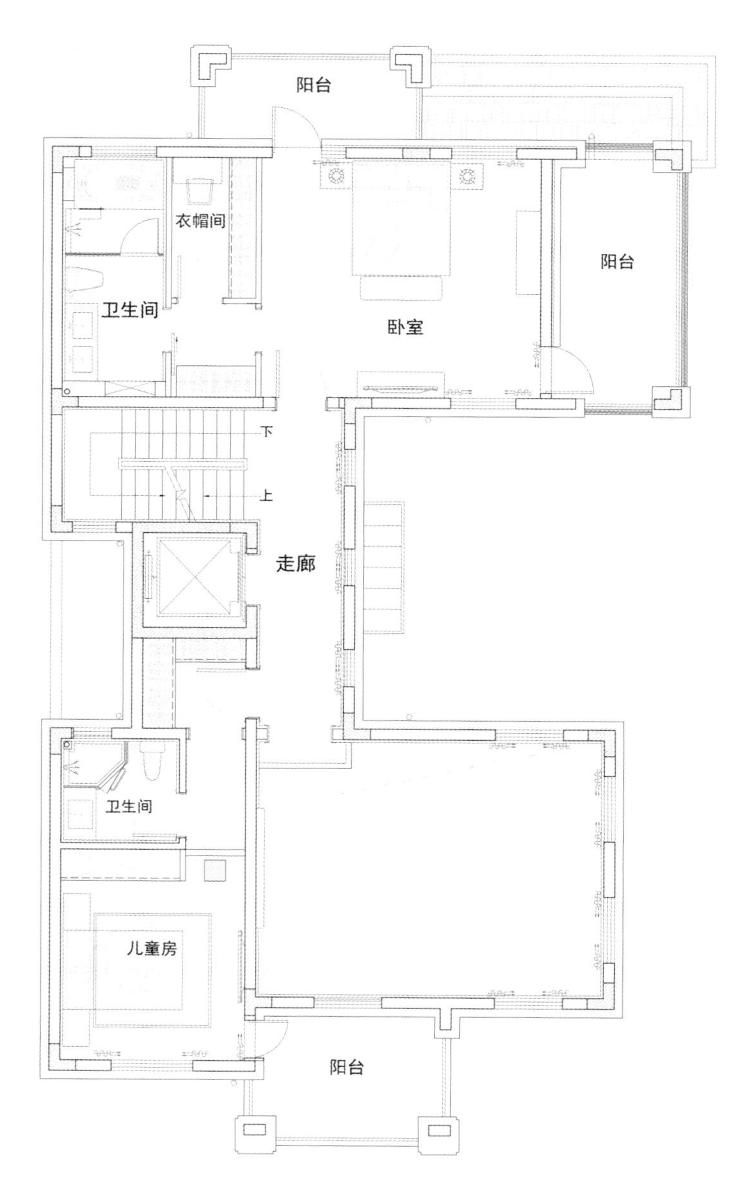
阳台
衣帽间
卫生间
卧室
阳台
下
上
走廊
卫生间
儿童房
阳台

Furniture color in the dining room is unified with the tone in the living room. Wood color dining table and gray cloth dining chairs are luxurious and fresh; concise and gorgeous droplight adds romantic atmosphere for dining time. Soft gauze curtain near the window blocks out the hustle and bustle of mortal world, making people feel tranquil and touching, adding a magnificent atmosphere for the living room and creating a plain fashion. Here is your independent space; home is not still any more but is endowed with the most sensitive meaning of life by the residents.

餐厅家具色彩与客厅总体的色调统一。木色的餐桌，灰色的布艺餐椅，奢华又不失清新，简约而不失华丽的吊灯，增添了用餐的浪漫气氛。窗边上轻柔的纱幔，将那凡世的喧嚣抵挡在外，给人的感觉是那样的宁静动人，也让客厅增添了大气感，营造出一个朴实之中的时尚。在这里，是属于你自己的独立空间，家不再是静止的，而是被居住者赋予了生活最感性的意义。

下沉庭院
窗井露台
储藏室
阳台
上
保姆房
卫生间
茶室
洗衣房
设备间
上
走廊
会客区
客卫
下
车库
棋牌室

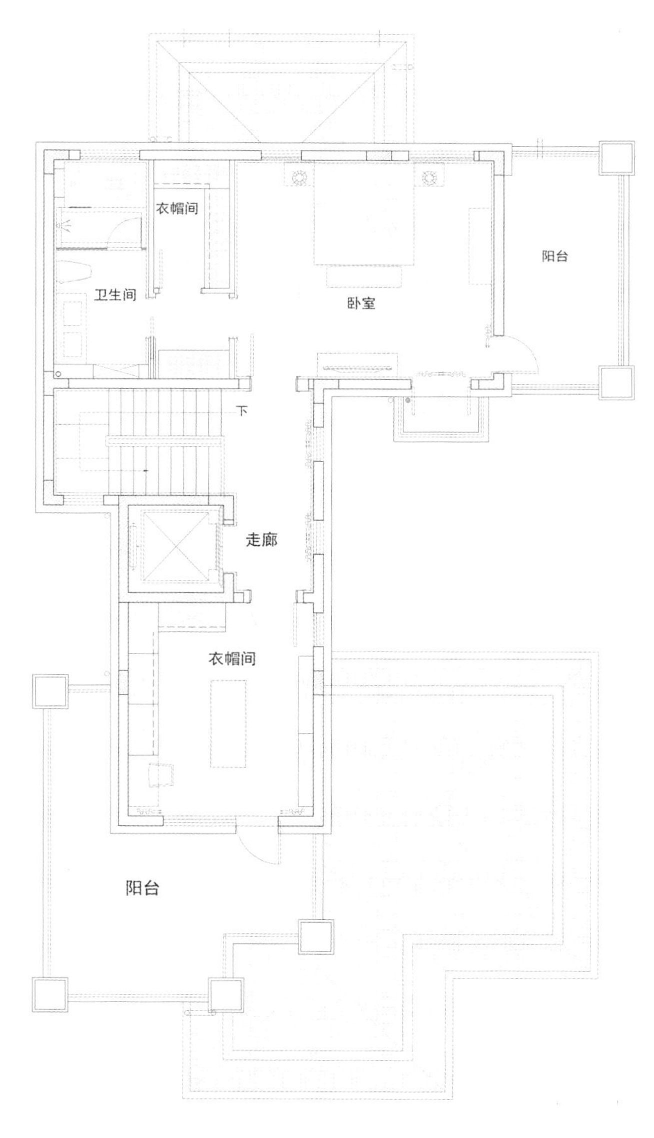
衣帽间
阳台
卫生间
卧室
下
走廊
衣帽间
阳台

PLEASANT AND GORGEOUS MANSION

悦居兰舍

Design Concept | 设计理念

This project is divided into two layers on the ground and one layer underground. Aiming at family structure of a family of four, the designer reorganizes space kinetonema to make the entire space well-regulated.

Considering the convenience of elders' daily life, the designer intentionally puts the elders' room in the first floor; the space is fluent and convenient for their daily lives. The most important living room manifests spacious and transparent. The second floor is rooms for the couple and their children, concerning that there should be more interactions and communications between parents and children and that the children are too young as that they need more concerns and cares from their parents. The penthouse design of the second floor particularly reflects the parents' profound love. The whole penthouse is children's recreation area; the parents can see their children's every movement when playing at the door of their room through the high glass barrier. As for the underground part, the designer turns the original submerged courtyard into a leisure area and SPA area with a sun roof, which fully divides the monotonous space into two. The leisure space becomes the space for the family of four to communicate with each other in their spare time and provides a place for friends to taste tea and discuss cheerfully and humorously.

项目名称：龙湾
设计公司：东易日盛别墅设计中心
设 计 师：金永生
项目地点：浙江杭州
项目面积：600㎡
主要材料：大理石、木地板、布艺、壁纸等

本案分为地上两层，地下一层。设计师针对整个四口之家的家庭结构，重新规划空间动线，使得整个空间变得更加井然有序。

考虑到老人生活起居的便捷性，设计师将老人的房间特意放到了一层，空间流畅、行动方便。而最重要的会客厅空间尽显了居室空间的空旷、通透。二层主要是男女主人和孩子的房间，主要是希望孩子和父母之间多些互动和交流，同时也是考虑到孩子年纪还小需要父母更多的关照和爱护。二层阁楼的设计尤为体现了父母爱的深厚：整个阁楼为孩子的游乐区，玻璃的高护栏在男女主人的房间门口都能看到孩子嬉笑玩耍的一举一动。地下一层部分，设计师将原有的下沉庭院改建成了带有阳光顶的休闲区和SPA区，充分地将单调的空间一分为二。休闲区成了四口之家闲暇之余彼此沟通增加亲情的空间，也使得和密友之间有了一个品茶论道、谈笑风生的必选之地。

The most interesting block stair of this project skillfully combines with glass guardrail; fluent metal lines connect every corner of the entire room, making the three generations link closely. In the whole design, the designer uses design concept of modern style and exquisite and vivid processing technique to create a warm, delicate and high-quality life atmosphere. At the same time, the designer positions this residence as a "time and space which inherits the past, feel now and carries the future". The comfortable modern space makes the elders live peacefully and harmoniously, the youth live happily and joyfully, the children grow up healthily, which is exactly the harmonious gesture which the space wants to present.

本案最有趣的积木楼梯与玻璃护栏巧妙结合，流动的金属线条联动着整个房间的每个角落，使之三代紧紧相连。整个设计中，设计师以现代风格的设计理念、细腻生动的处理手法，打造温馨、精致和高品质的生活氛围。同时也把该住宅定位为一个“传承过去、感受现在、承载未来的时空”，在舒适的现代空间中让老人其乐融融、颐养天年，年轻人乐得其所，孩子健康成长，正是该空间所要呈现的一种和谐姿态。

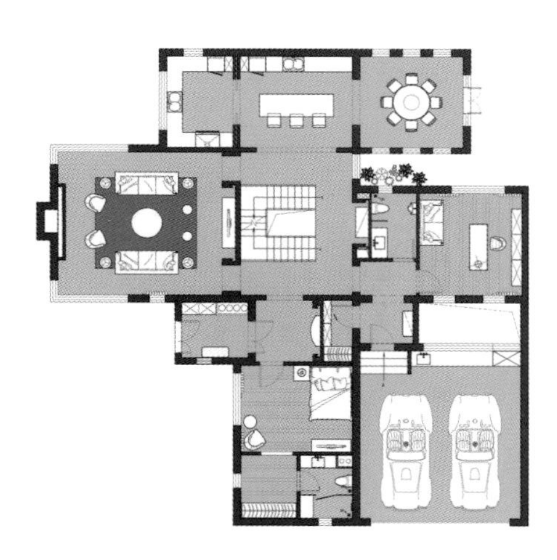

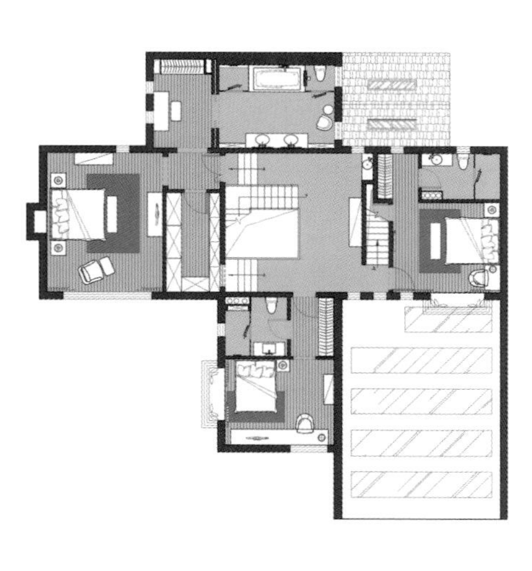

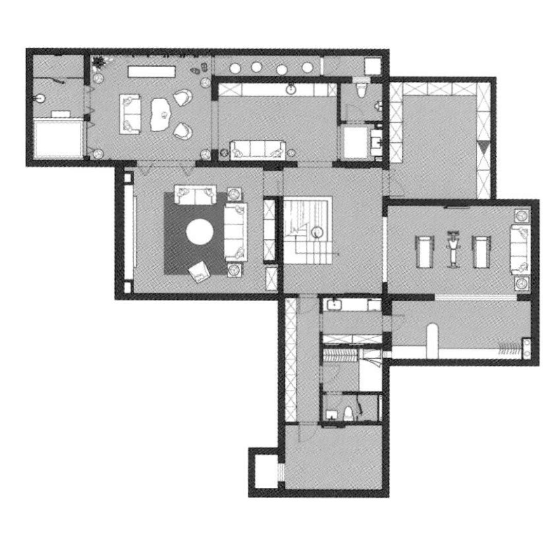

LUXURIOUS ROMANCE

奢华的浪漫

Design Concept | 设计理念

The style of this project is French style. The space is filled with fresh and romantic atmosphere, retro, luxurious and natural feeling. The living room sets warm beige as basic tone, collocates with light elegant and plain gray, fresh and lively blue and clam brown and forms a romantic and fresh French space. As for decorations, baroque wood carvings, bright crystal droplight, various furnishings, green curly grass pattern curtains and full flowers reveal heavy retro feelings. The dining room follows the classic French style; the surfaces of tables and chairs have carvings, with arcs of chair backs and legs, manifesting elegance and nobility; round ceiling design highlights romantic and lively temperament of the space. In the bedroom, delicate and gorgeous beddings match with entire pure environment, fresh and elegant, manifesting French elegance and nobility. This kind of "nobility" emits humanistic and classical flavors; comfort, elegance and easiness are its inner temperaments.

本项目定位为法式风格，空间中弥漫着清新、浪漫的氛围，充满了复古、奢华、自然主义的情调。客厅以温暖宜人的米色为基础色调，搭配淡雅素净的灰、清新活泼的蓝、稳重的褐，形成浪漫清新的法式空间。装饰上，巴洛克式的木质雕花、璀璨的水晶吊灯、琳琅满目的摆设、青色的卷草纹窗帘、饱满的花艺搭配等，透露出浓郁的复古风情。餐厅秉承经典的法式搭配风格，桌椅表面略带雕花，椅背和椅脚的弧度，显得优雅矜贵，圆形吊顶的设计则更突出空间浪漫灵动的气质。卧室中，精美华丽的床品，配上整体纯净洁白的环境，清淡优雅，尽显法式的高雅矜贵。这种"贵"散发着人文和古典气息，舒适、优雅、安逸，是它的内在气质。

项目名称：山东雪野湖桃花源什锦园法式C1户型
设计公司：北京中合深美装饰工程设计有限公司
设 计 师：张杰
项目地点：山东莱芜
项目面积：248m²
主要材料：壁纸、实木、大理石等

RETURN TO THE ORIGIN OF HOME

回归家的原点

Design Concept | 设计理念

When a designer has been weary of concept stacking, and no longer wants to argue for his own design proposal by pompous excuse, instead, he hopes to manifest his unique thinking of space and life in the design, how can he do?

For excellent home design, designers always treat the house as a "home", its unique sense of belonging would arouse far more sympathetic chord than the "style" and the "principle". It should be warm and has sentiment. For each family member, it is not only a beautiful and warm place, but also a safe harbor in their long lives forever. In this case, the villa design is a kind of interpretation of this thinking.

当一位设计师厌倦了概念的堆砌，不再想以冠冕堂皇的说辞来为自己的方案辩解，转而希望在设计中真正体现自身对于空间与生活本质的思考，他会怎样做呢？

对于优秀的家装设计而言，住宅作为"家"所特有的归属感可能远比"风格"和"主义"更能引起人们的共鸣。它应该是温暖的，有感情的，对于家庭中的每一位成员而言，这里是美丽而温馨的归宿，是漫漫人生旅途中永远为你预留的避风港湾。本案中这栋地上两层，地下一层的别墅设计，正是设计师对于这番思考的一种诠释与演绎。

项目名称：杭州桃花源沈宅
设计公司：PMG.伍重室内设计
设 计 师：梁苏杭
项目地点：浙江杭州
主要材料：大理石、木地板、布艺、壁纸等

Walking into this beautiful mansion, you can first see the delicate welcoming foyer. The two-story ceiling brings plenty of lights. When seeing the pastel colors, vintage parquet floor and beautiful armchairs with soft lines, visitors would immediately pull themselves out of the gray and fast-paced modern environment, and back to the 20th century to enjoy the elegant life of middle class; the dark blue carpet which paves over the beautiful stairs straight to the second floor and the black iron handrail together outline the space transition and stack; the blue peacock specimen stands on the viewing platform on the second floor and looks like a pretty hostess in a long skirt that would step down the stairs in an elegant gait at any time. What a fantastic imagination.

Entering the house through the foyer, you can see the dining room and kitchen on the left of the corridor. As an important place of modern people to hold social and family gatherings, the dining room often reflects the owner's interest in life. Here, the designer has continued the foyer's soft and bright color tone. The jasper wall echoes with dining chairs, creating an enthusiastic and luxurious European style. The opening western kitchen and the closed Chinese kitchen which respects the Chinese tradition will provide more options for family gatherings and hospitality. At the same time, it would also prevent the flue gas into the indoor when cooking in high temperature.

移步进入这栋美丽的大宅，首先看到的是精致的迎客门厅。两层挑高带来了充足的采光，柔和的色彩、复古的石材拼花地面和线条柔美的扶手椅，将步入门厅的客人迅速从灰色调、快节奏的现代环境中抽离出来，带回到20世纪中产阶级优雅的生活中去；那直通二层的美丽楼梯所铺设的深蓝色地毯，与黑色的铁艺扶手一同勾勒出空间的转折与层叠关系；二层观景台上矗立着的蓝色孔雀标本，则犹如身着长裙的美丽女主人随时会以优雅的步态迈下台阶，令人浮想联翩。

通过门厅步入宅邸，走廊的左手是餐厅和厨房。作为现代人重要的社交及家庭聚会场所，餐厅风格往往体现着主人的生活情趣。在这里，设计师延续了门厅柔和明亮的色彩基调，墨绿色的墙面和餐椅相呼应，为餐厅营造出奢华热情的欧式风情。开放的西式厨房和封闭的中式厨房尊重中国传统，为家族聚会及待客提供了更多选择，同时也保障了高温烹饪过程中的烟气不至影响室内环境。

On the right hand, it is the living room. Compared with the dining room, it creates a more pleasant and relaxing atmosphere. Gray wall decoration and sofa make people feel relaxed. Adequate day lighting from two French windows of heavy drapes on both sides can meet the master's demands to switch the living room scenes at any time. It takes both transparency and privacy into account.

右手是客厅，相对于餐厅来说，客厅营造的是一种更加松弛和愉悦的气氛。灰色调的墙饰和沙发，让人有一种柔软的放松感。两面落地窗充足的采光配合厚重的布帘，让客厅的场景可以配合主人的需求随时切换，兼顾通透与私密两种要求。

The study hides behind the living room. Its subdued colors and concise decorations make people calm down. And it links with the sunlight room through the French windows so that the owner can open the curtains and enjoy the natural scenery any time he feels tired. When cold winter comes, lighting the fireplace in the middle of the study, sitting on the armchair beside and reading a good book, everything is no doubt a wonderful enjoyment for the owner.

书房掩藏在客厅的后面，色彩沉稳，装饰简洁，令人沉静。同时这里通过落地窗与阳光房相连，使得主人在工作疲劳时随时可以打开窗帘，欣赏窗外自然风光。寒冷的冬夜点燃书房中间的壁炉，坐在旁边的扶手椅上听着炉火噼啪，坐拥一本好书，无疑也是人生一大享受。

Back to the foyer and up to the second floor along the stairs, we arrive to the more private areas, the owners' living area, including the master bedroom, nursery and children's bedrooms.

If the study reflects the host's interest, then the master bedroom embodies the romanticism of the hostess. Warm and soft colors, natural decorations, floral and lace fabrics, soft carpet and bed chair, all of these create a warm and romantic atmosphere. Family photos are visible in this room, always reminding us that the family live a happy life.

退回门厅沿着楼梯走上二楼，则到达了宅邸更私密的区域——主人们的起居区，主要包含主卧、育婴室、以及孩子们的卧室。

如果说书房反映的是男主人的爱好，那么主卧则是女主人的情怀。温暖柔和的色彩，自然主题的陈设装饰，点缀花边和蕾丝的布艺，柔软的地毯和床尾凳，无一不营造着温馨浪漫的氛围。房间内随处可见的家族照片，随时提醒着我们主人家生活的幸福美满。

The birth of baby is a symbol of new life and is also a continuation of a family. White walls with lath decoration and slope roof in the nursery create a romantic French style. Small tables, chairs and sofas in the model of pumpkin carriage lead the princes and princesses into the scene of fairy tales. Childish decoration will keep their childhoods forever. In addition, with children growing up rapidly, they would leave to their own rooms. The boy's room is hale and firm while girl's room is romantic and warm. I believe, children who leave from here must have their own splendid lives.

婴儿的诞生是生活开启新希望的象征，是家庭梦想的第二次启程。育婴室的白色板条装饰墙面和坡屋顶营造出浪漫的法式情调。小小的桌椅和沙发犹如南瓜马车一样的造型，将王子和公主们带入迷人的童话场景。充满童趣的装饰品将孩子们的童年永远留在了这里。此外，孩子们的成长总是令人惊叹，两位离开育婴室的小主人们也拥有自己的房间：男孩房硬朗坚毅，女孩房浪漫温馨，从这里离巢的孩子们一定也会拥有自己精彩的人生。

In other words, the construction of a house not only embodies the hard work and wisdom of the owner and designer, but also carries their hopes and understandings for lives. The completion of the construction does not mean the completion of the project since the residents will bring energies and colors into this house and make it more perfect; on the contrary, the house itself will provide much more wonderful memories and joys to them. It is the common growth of house and residents that builds life itself.

或许可以这样说：一栋住宅的建造，倾注着主人和设计师大量的心血，同时也寄托着对于生活的理解和希望。工程的完工并不意味着项目的完成，居住者会在今后的生活中继续为这栋住宅增添色彩，使它日臻完美；而住宅本身，也会回报给居住者更多美好的回忆和享受。正是住宅与居住者的这种共同生长，构建了生活本身。

FASHIONABLE AND CLASSIC BALANCE AESTHETICS

时尚与经典的平衡美学

Design Concept | 设计理念

Someone once said, life is a journey and the passed landscape is the scenery; the experienced joy is the happiness. No matter prosperity or wateriness, all is your own scenery, please go on quietly and indifferently. This kind of indifferent state is conveyed into the space. ART DECO furniture combines with Chinese flower arrangement; concise construction, elegant temperament and classical beauty coincide naturally and present a unique and harmonious aesthetics. As for colors, sunshine orange is used throughout the elegant gray interior space, with strong red, pure blue and full yellow, creating comfortable living atmosphere. Living here, one can see the fleeting fireworks and taste tranquil time.

有人说，人生就是一次旅行，走过的山水，都是风景；尝过的欢愉，都是幸福。无论好与不好，无论繁华与平淡，都是属于自己的风景，还请你安静淡然地走下去。这种泰然自若的心态也传递到空间中：ART DECO风格的家具与中式插花相结合，简单的构造、典雅的气质与东方的古典之美不谋而和，呈现出独特和谐的美学。色彩上，阳光般的橙色穿行于优雅的灰色调室内空间中，穿插浓艳的红、纯粹的蓝、饱满的黄，深浅浓淡总相宜，营造出惬意的生活气氛。徜徉于此，淡看流年烟火，细品岁月静好。

项目名称：北京密云叠拼三单元户型
设计公司：北京中合深美装饰工程设计有限公司
设 计 师：郭小雨、石哲
项目地点：北京
项目面积：850㎡
主要材料：大理石、木地板、布艺、壁纸等

The designer links the living room with dining room, which visually enlarges size of the space and virtually strengthens the flowing sense. Classic fret elements, hollowed-out screen, Chinese lamp and ink paintings with full imagination bring viewers multi-level visual feelings. In the bedroom, plain and elegant flower and bird background wall is warm and natural as if in the sweet and sunny spring, leisure and free. The bedside lamp emits quiet and gentle lights; gray brown beddings make people deposit unconsciously.

设计师将客厅与餐厅放置在一起，视觉上扩大空间尺寸感的同时，流动感也无形中被加强。经典的回纹元素、镂空间隔屏风、中式台灯、让人产生丰富遐想的水墨画等，带给观者多层次的视觉感受。卧室里素雅的花鸟背景墙温馨自然，如同置身于甜美明媚的春光里，慵懒自在。床头灯散发出静静的温柔的光芒，灰褐色的床品观之让人心绪不自觉沉淀。

SAFARI
SERENGETI
KELLY HOPPEN STYLE

NEO-
STYL

CHINESE

新中式风格

CONTEMPORARY ORIENTAL NEW HUMANISTIC LIFE

当代东方新人文生活

Design Concept | 设计理念

Bamboo sprout is used to catch fish and when the fish is caught, the sprout is forgotten; words are used to convey meanings and when the meanings are understood, the words are forgotten.

—— *Waiwu, Zhuangzi*

Ocean Palace is located in the Fifth Ring Road in Beijing with cultural deposits of six dynasties and ancient capitals flowing in its blood as well as international lights lighting up its window, so it must be Oriental and it needs modern life feelings. Here LSDCASA digs kind beliefs which fit the contemporary universal value and spirits to express Oriental culture, turns them into contemporary forms by our crafts and manifests "Orient" in spiritual and cultural identity. Therefore, the whole space has no significant Oriental symbol piles and reveals Oriental Zen tranquility all the time.

筌者所以在鱼，得鱼而忘筌；
言者所以在意，得意而忘言。
——《庄子・外物》

远洋天著位于北京五环，六朝古都的文化沉淀，流淌在它的血液当中，而国际都会的灯光亦点亮它的窗台，它必然是东方的，但它也要有现代的生活感受。LSDCASA在此处，对东方文化的表达选择通过挖掘传统中良善的、符合当今普世价值的信仰、精神，用我们的技艺转化为当下的形式，以精神、文化层面的认同来对切"东方"。因此，在整个空间中没有任何显著的东方符号堆砌，却自始至终浸透着东方特有的禅思静谧。

项目名称：北京远洋天著平墅
软装设计：LSDCASA
设计团队：LSDCASA设计一部
项目地点：北京
项目面积：362㎡

The general parts manifest hardness and the details manifest softness; all manifest the elegance. The sun inclines and sparkles a tranquility for the whole space. The pure silky carpet paves the space atmosphere; the couch comfortably lies in the central; behind it is the family's favorite things; when guests come, there are many memorable stories to share; the friendly living room is full of culture breath.

The tea table modeling is chic, half of which is made of marble natural texture to form ink painting images while the other half is neat lines to present modern industrial appearance, which exactly fits our understanding and starting point of the space. BoConcept floor lamp comes out from the back book shelf and integrates into the French window scenery; seeing from a distance, it seems the northland branches, activating layering of the whole living room.

大处见刚，细部现柔，不着一笔而尽得风流。日光半斜，为整个空间洒下静谧。丝质的纯色地毯，铺满空间的氛围，长沙发舒适地安放在中央，背后陈列着一家人乐于展示的心水之物，每当有客来访，如数家珍的故事桩桩件件，友好的客厅瞬间弥漫文化气息。

茶几造型别致，一半由大理石的自然肌理构成水墨意象，一半由工整的线条写就现代工业风貌，正契合我们对这个空间的理解和出发点。BoConcept的落地灯自后排书架斜出，融入落地窗景，远远望去，好似北国的树枝，生动了整个客厅的层次。

In the spacious dining and kitchen space near the living room, what presents in your eyes is the hand-painted colorful floor screen; the warm color walnut dining table and chairs make the dining space more independent. Italian SELETTI solar system planet porcelain cup and pure copper meal buckle match with black cotton and linen napkins, as if the long night and bright moon, echoing with the ancient Chinese round window artistic conception.

When the night falls, you can walk through the gallery of corridor and enter into the bedroom. Lighting the lightening copper lamp, warm lights evenly sparkle on the silky beddings which softly glow lights. the decorative cabinet is put with flowers of the wife and tea set bought from Hong Kong. The wife is dressing up near the dresser and preparing to go to the party with the husband; The husband sits on the sofa and quietly enjoys; time flows between the familiar tacit understanding.

客厅一侧开敞的餐厨空间，首先映入眼帘的是手绘韵彩落地屏风，对映暖色系的胡桃木餐桌椅，使用餐空间更加独立。来自意大利SELETTI太阳系行星瓷盘，纯铜餐扣搭配黑色棉麻餐巾，好似长夜月明，又似遥遥呼应了古代中国圆窗的意境。

暮色将近，走过画廊式的过道，进入到主卧。点亮Artemis铜质台灯，暖色的灯光均匀地洒在丝质的床品上，柔和地泛着光。装饰柜上端放着太太摆弄的花艺和从香港淘回来的茶具。太太正在梳妆台的一处打扮着，准备与先生参加聚会。先生坐在一旁的沙发上静静的欣赏着，时光在熟悉的默契间流动。

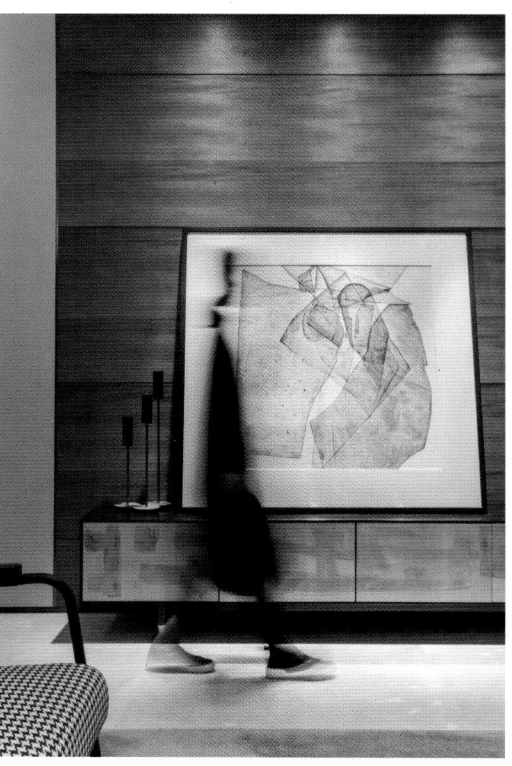
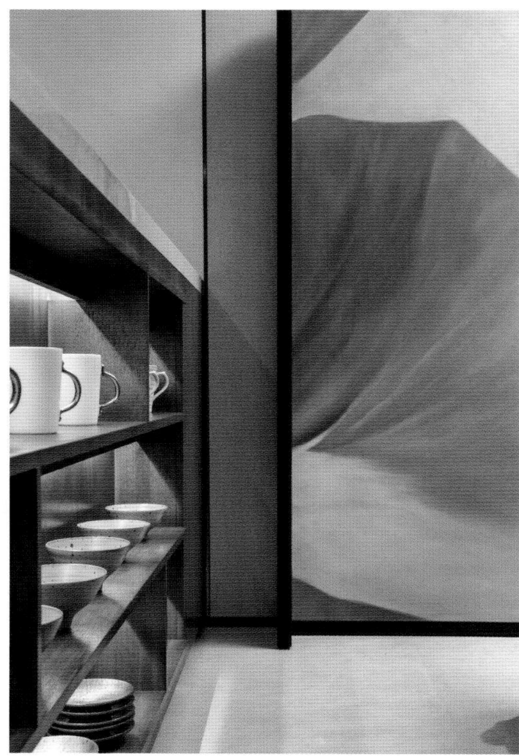

THE PAIN YOU FEEL TODAY
WILL BE THE
STRENGTH
YOU FEEL TOMORROW

The end of the corridor is the boy's room; in the lively blue and bright yellow main tone, his vivaciousness and optimism are revealed totally. Art geometric segmented painting echoes with the carpet. In the ancient time, flower arrangement, painting, tea and incense were called "four arts of life", and here it is the way for the hostess to release love to art. The art room connects with the lower yard with a mulberry tree; under the tree is a wood tea table with native rattan meditation mat, as if you can press the pause key of time and have a half day leisure.

Chinese traditional flower arrangement is the origin of Oriental flower arrangement art. In the process of arranging flowers, the person needs to "cut, give up and discard". So does human. Through cutting, giving up and discarding, people empty environment and distracting thoughts and live a simple life so as to welcome the free and comfortable life. So does design. A good designer is nothing but better at choosing, not to choose what he wants but to choose what he doesn't want. This space narrates this point of view by extreme reset and subtract.

过道的另外一头就是男孩房，明快的蓝色、亮黄奠定空间的主色调，他的活泼、开朗在此已展露无遗。艺术几何分割的挂画，与地毯相互呼应。在古代，插花、挂画、点茶、燃香被统称为“生活四艺”，而在这里只是女主人释放对艺术的热爱的方式。研艺室交接着下层庭院，庭院里种着一颗桑果树，树下是木质做成的茶桌，原生态藤编质的禅修垫，仿佛能在这里按下时间的暂停键，偷得半日浮闲。

中国传统插花是东方插花艺术的起源，在插花的过程中，时常需要花艺者“断舍离”。人亦如此，通过断舍离，人们清空环境，清空杂念，过简单的生活，才迎来自由舒适的人生。设计更是如此，一个好的设计师无非比他人更擅长选择——不是选择要什么，而是选择不要什么。这个空间以极致的清减，叙述这一观点。

CATCHING A GLIMPSE OF ELEGANCY IN CROWD

喧嚣处寻觅逸趣风雅

Design Concept | 设计理念

This elegant building by the lake, integrated into the landscape, will never fail to fascinate people. The emotion inspired by the atmosphere of elegancy, making people who live here enjoy the serenity of nature in the busy urbanism. The theory of harmony founded in the design concept. Black and white decorations create the balance from the aesthetical point of view. Hundreds splendid golden stars radiate dazzlingly underneath the marvelous dome. Glamour vases are set on the Ming style furniture nearby the fly-away plants. All these settings are appreciated by those sophisticated citizens.

傍水而居，和山水融为一体，在喧嚣的闹市中清幽自得，在含蓄节制的空间中屏蔽干扰，心灵真正的得以升华。是白，是黑，是节制，也是浮夸；是静，是动，是和谐，也是矛盾。6米高的挑空玄关，泼洒下1000多颗璀璨的金属雨滴。流韵点金是浮夸的，更是震撼的。《见南山》边柜上中式摆物是寂静的；红色花器里悄然窜起的龙柳是躁动的……这深深浅浅的矛盾与道要每一位缘者细细的品鉴。

项目名称：翡翠松山湖---滨湖花园4D#01户型

设计公司：深圳市圣易文设计事务所有限公司

项目地点：广东东莞

项目面积：1000m²

主要材料：翡翠木纹大理石、水晶白大理石、蜘蛛玉大理石、红龙玉大理石、梅尔斯金大理石、木饰面等

Over here, philosophy of East and technology of West make an interesting combination which brings unique DNA to the living room. Furniture has been settled at the most appropriate position. Grass and trees outside the window bring the inner peace to every guest in the saloon. Sun lights, as the best morning ring for a good day, wake people up at dawn and make their dream seemingly come true.

围合式的家庭厅带着独特气韵和生命灵性。中式禅意与西洋玩物的碰撞，逸趣横生；敦实的布艺沙发与纤细的传世家具相得益彰；天花的艺术吊灯与高低错落的茶几相会呼应，如同休止符，欲言又止。将视野延伸到窗景外，满眼的草木绿意，宁静与虚实间塑造出清雅意境。流年似水，静候的每个寂静清晨，被倾洒进来的晨光唤醒。墨色似烟雾缭绕，被阳光拨开，无需多余的笔墨，缘者便能了悟当中的意境。

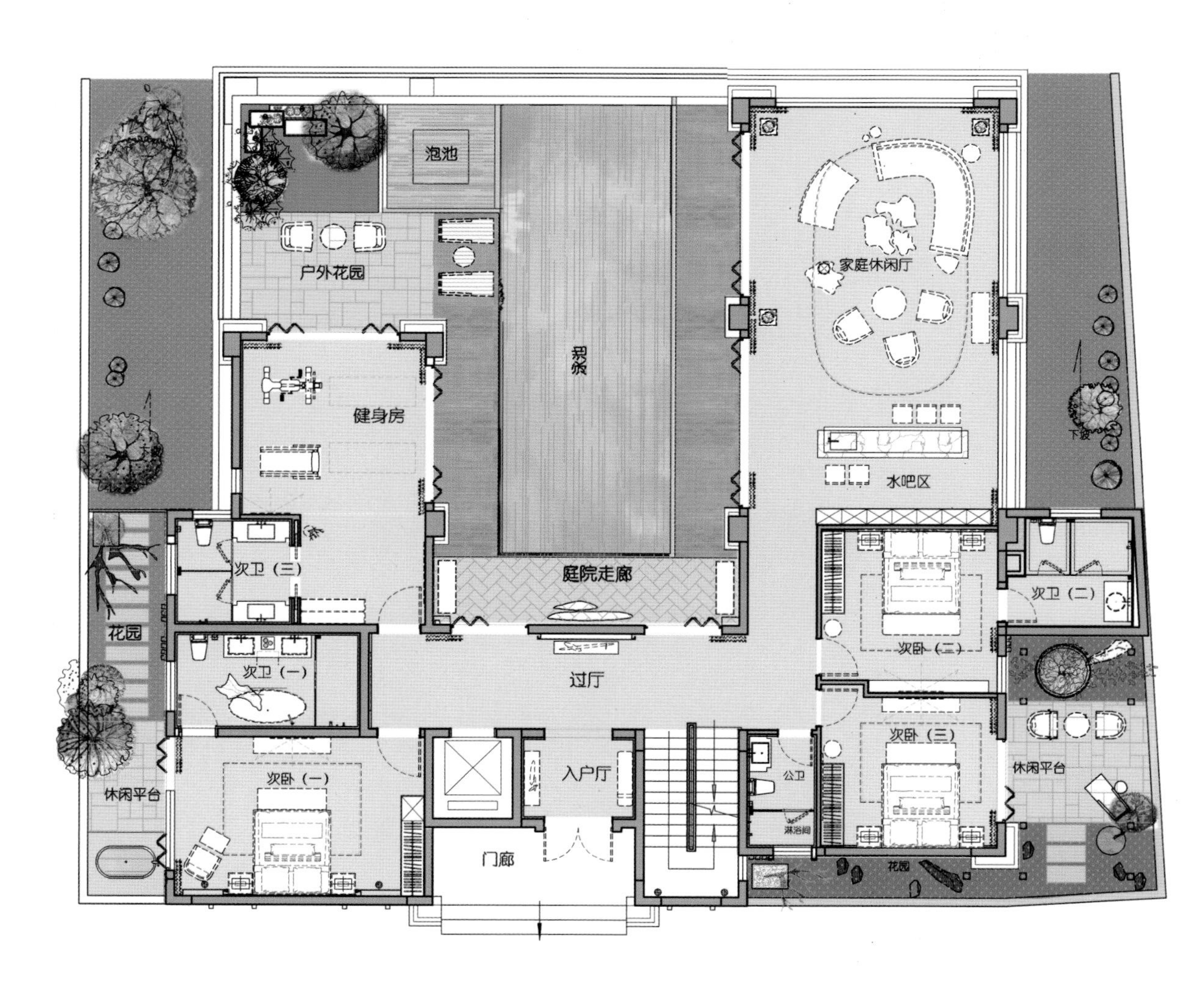

泡池
户外花园
家庭休闲厅
健身房
水吧区
次卫（三）
庭院走廊
次卫（二）
花园
次卧（二）
次卫（一）
过厅
次卧（三）
入户厅
公卫
休闲平台
次卧（一）
休闲平台
门廊
花园

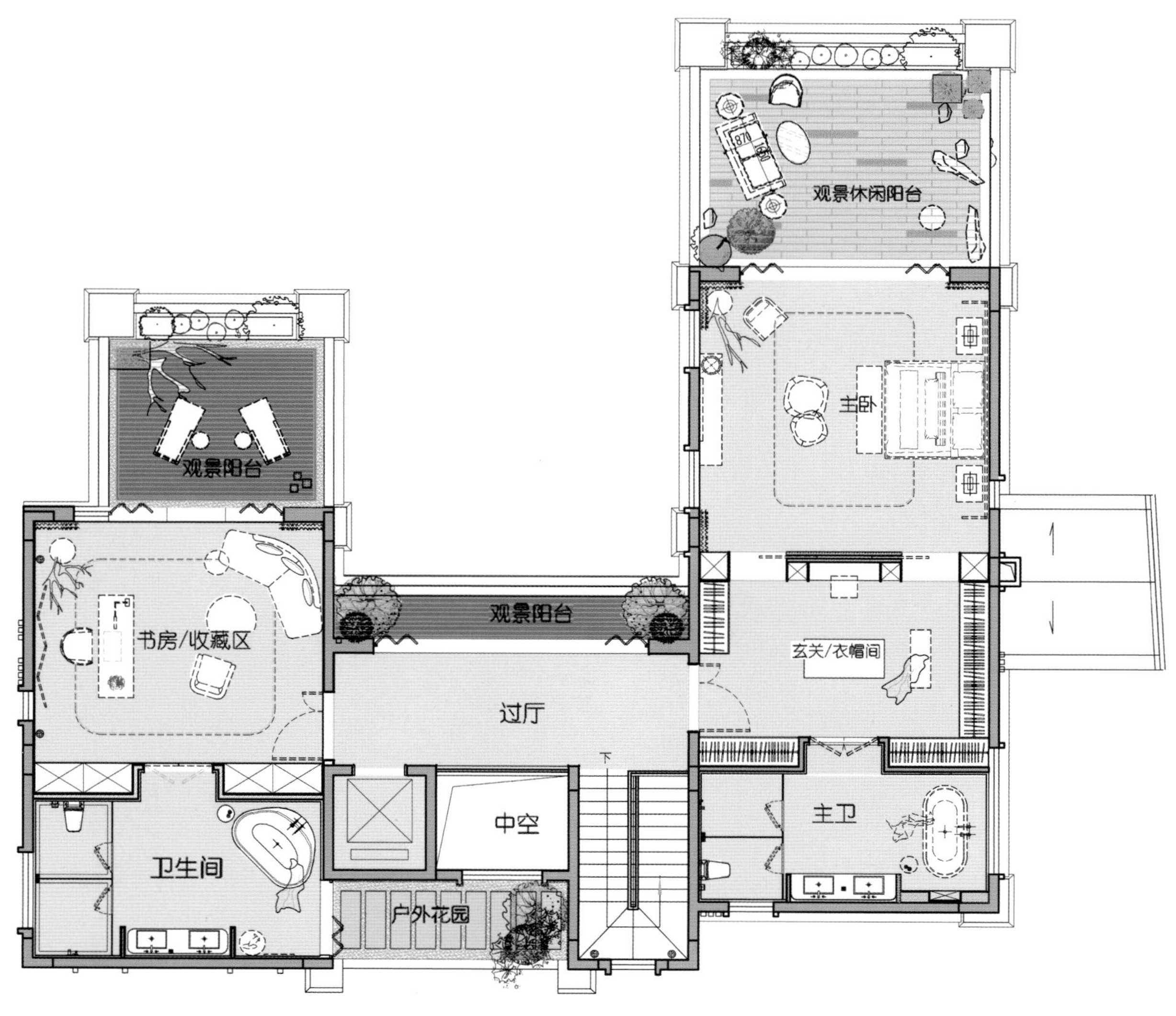

观景休闲阳台
主卧
观景阳台
书房/收藏区
观景阳台
玄关/衣帽间
过厅
下
中空
主卫
卫生间
户外花园

Balcony is always the best place for talking. The company of friends could warm people even in the rainy winter night. The yard is full of spiritual energy with such good view. People could always enjoy a peaceful moment here. The interior decoration features traditional Chinese painting which reveals a sense of ancient Chinese favor.

雪梅化作白色的涟漪，漂浮起舞在水面上。缘者揣一颗平常心，从容淡然地闲庭信步，时而看世间风轻雨淡，时而三五知己高谈理想。风与声的交汇，青葱绿意，潺潺流水，平静中蕴含着空灵。与风景融为一体，感受精神的力量。一剪寒枝，期待与缘者的一次邂逅。笔墨深浅，寂寥无声，运用留白艺术，将空白延伸扩展，勾勒山水意境。

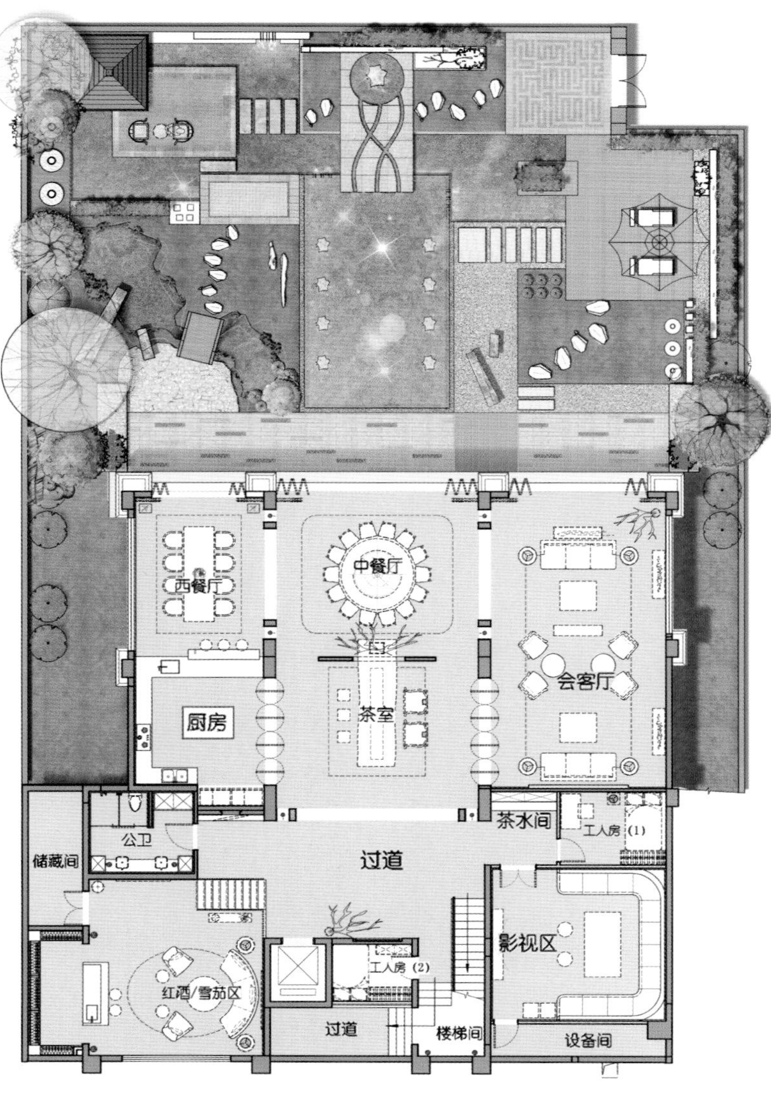

西餐厅
中餐厅
厨房
茶室
会客厅
公卫
茶水间
工人房（1）
储藏间
过道
影视区
工人房（2）
红酒/雪茄区
过道
楼梯间
设备间

The usage of Chinese Screen splits the space, also reunites them with traditional Eastern aesthetic by means of the painting on the Screen. It is a bold design to apply a symmetrical layout for the parlor. The ceiling with a specific pattern, the lantern in concise style, the match of lake blue and sunset orange, the furniture covered by traditional pattern texture, all those elements in the parlor present a Chinese favor through western method, wishing to let people catch a glimpse of elegancy in crowd.

运用构成的设计手法，屏风分隔了两个空间，同时又融合了它们，是动，是静，是矛盾。绢布上若隐若现的红梅与茶桌上挑出的罗汉松，是对比，又是和谐。在如此一个充满了矛盾的和谐空间里，寻找心之所向。对称式布局的会客厅融汇了设计师的大胆塑造。复杂而隆重的天花纹样、简洁现代的艺术云灯；湖水蓝与落日橙的搭配；中式纹理搭配现代的家具等等。用不拘一格的西方审美营造静谧的东方美学。正如在喧嚣处寻觅逸趣风雅，怡然自得。

INHERITING ORIENTAL AESTHETICS

传承东方美学

Design Concept | 设计理念

The real voyage of discovery consists not in seeking new landscapes but in having new eyes.

——*Remembrance of Things Past* by Marcel Proust

Design has been looking forward to innovation. In terms of Oriental aesthetics, we pursue classic not classical; we pursue trend rather than transitory popularity. The understanding of Chinese culture no longer stays in gilded and red appearance of carved dragons and painted phoenix but in digging life aesthetics of ancient refined scholars, seeking the ancient and modern balance under modern lifestyle and integrating the Oriental connotation and poetry of "her face half-hidden behind the lute in her arms" into life.

真正的发现之旅，不在于发现新的领域，而在于拥有新的眼光。——《追忆逝水年华》作者马塞尔·普鲁斯特

设计一直期待着创新，仅就东方美学而言，我们追求经典而非古典，我们追求潮流而非昙花一谢的流行，对中国文化的解读不再停留于雕龙画凤、描金抹红的表象，而当深掘古代文人雅士的生活美学，寻找现代生活方式下古与今的平衡，让东方“犹抱琵琶半遮面”的含蓄与诗意之美融入生活。

项目名称：珠海西湖湿地国际花园A4别墅
设计公司：深圳高文安设计有限公司
设 计 师：高文安
项目地点：广东珠海
项目面积：420㎡
摄 影 师：KKD推广部
主要材料：山水玉石、大理石、橡木、钢化清玻璃等

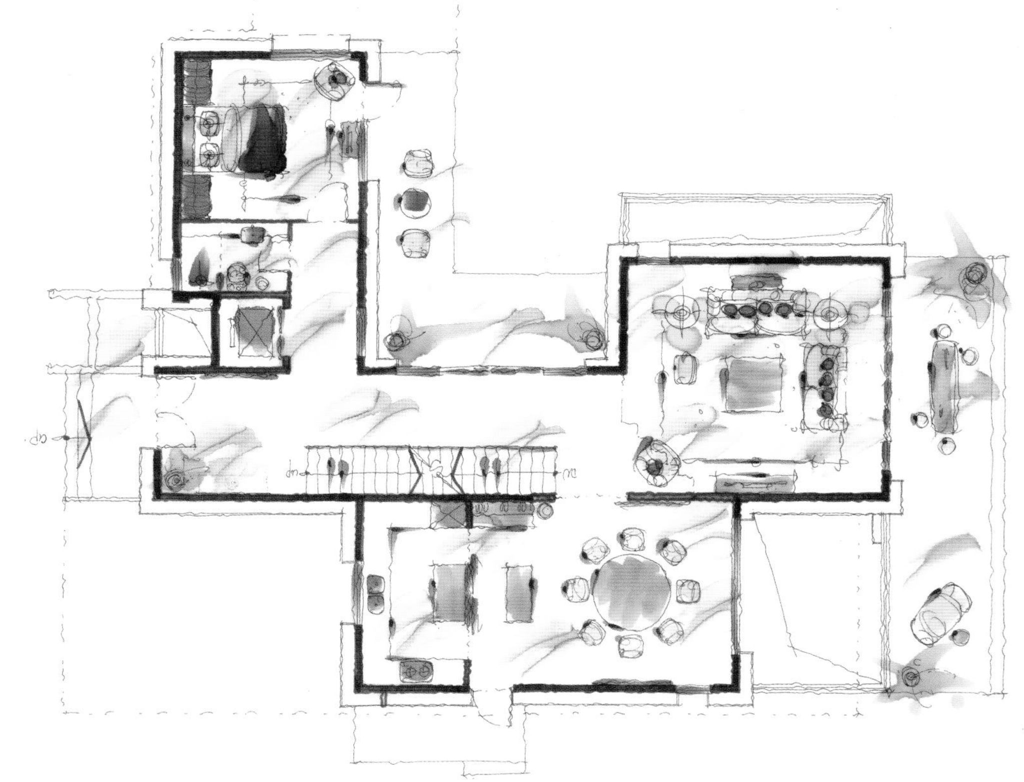

Sticking to the lifetime design is the goal which Kenneth Ko has been pursuing. He thinks that when completing his own works, a good designer must play three roles, the arbiter between ideality and reality, inheritor between the past and the future and the disseminator between the beauty of image and the beauty of object.

Interior design in the end is to design a life. As one of the top designers and living experts, Kenneth Ko knows that besides the unchanged technique, the power that culture and story endow to space and home is the life aesthetics which can withstand the test of time. Home needs to have the smell of home, the milk of human kindness, the taste of culture and the fun of life. The indifferent world is outside while the warm story is inside.

坚守一生的设计，是高文安孜孜不倦追求的目标。他认为一个好的设计师，在完成自己的作品时，必须扮演好三类角色：理想与现实之间的仲裁者，过去与未来之间的传承者，意象之美与物象之美的传播者。

室内设计说到底是在设计一种生活，作为顶级的设计大家和生活家，高文安深谙在万变不离其宗的技法之外，文化和故事所赋予空间，赋予家的力量，那才是经得起时间流逝的生活之美，家要有家味，有人情味，有文化品味，有生活趣味，冷漠的世界在外头，温暖的故事在里头。

The ancient Roman philosopher Cicero said: "all natural things are beautiful."

The design of living room pursues the natural effect of "moistening everything silently"; life temperature weakens design traces and strengthens cultural connotations in indifferent exquisiteness. As the poet Zhang Daoqia from Song Dynasty described in the poem *Plum Blossom in Dayu Ridge*: everywhere is picturesque; everything can be artistic conception in ancient poetry; beautiful things appear at an time. The plum blossom is all over the mountains and plains so that it attracts many visitors. Refining traditional elements is closely linked to the theme of "gentleman as jade". The plum blossom wall painting from one of the four gentlemen plum blossoms, orchid, bamboo and chrysanthemum uses landscape jade to create TV background wall; Chinese sensibility deduces modern artistic essences and fulfills the sublimation of art and culture.

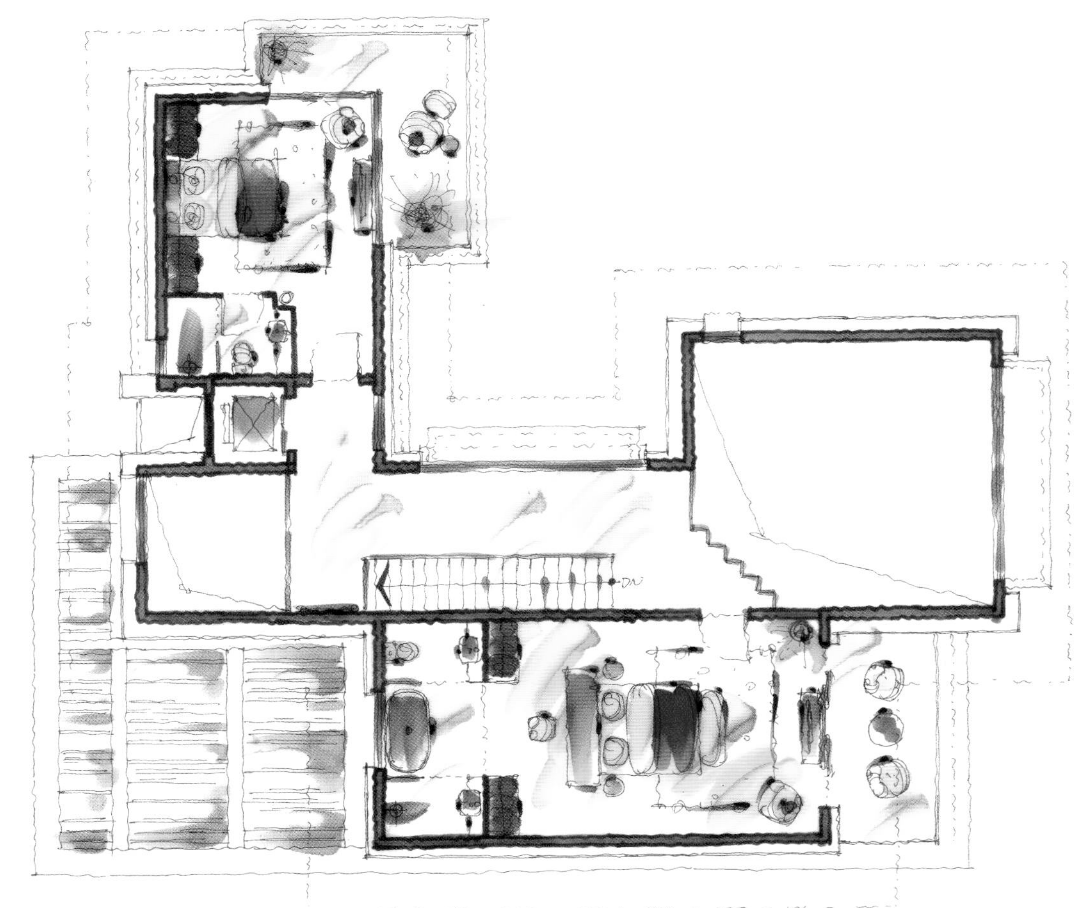

古罗马哲学家西塞罗说：“一切顺乎自然的东西，都是美好的。”

客厅的设计更多追求的是“润物细无声”的自然效果，让生活的温度淡化设计的痕迹，在淡然的精致中迭加深厚的文化底蕴，正如宋张道洽《岭梅》一诗中所写：到处皆诗境，随时有物华。应酬应不假，一岭是梅花。对传统元素的提炼，是紧扣“君子如玉”这一主题，“梅兰竹菊”四君子之一的梅花壁画，以山水玉打造的电视背景墙，皆以中式的感性演绎极具现代感的艺术真髓，完成艺术与文化的升华。

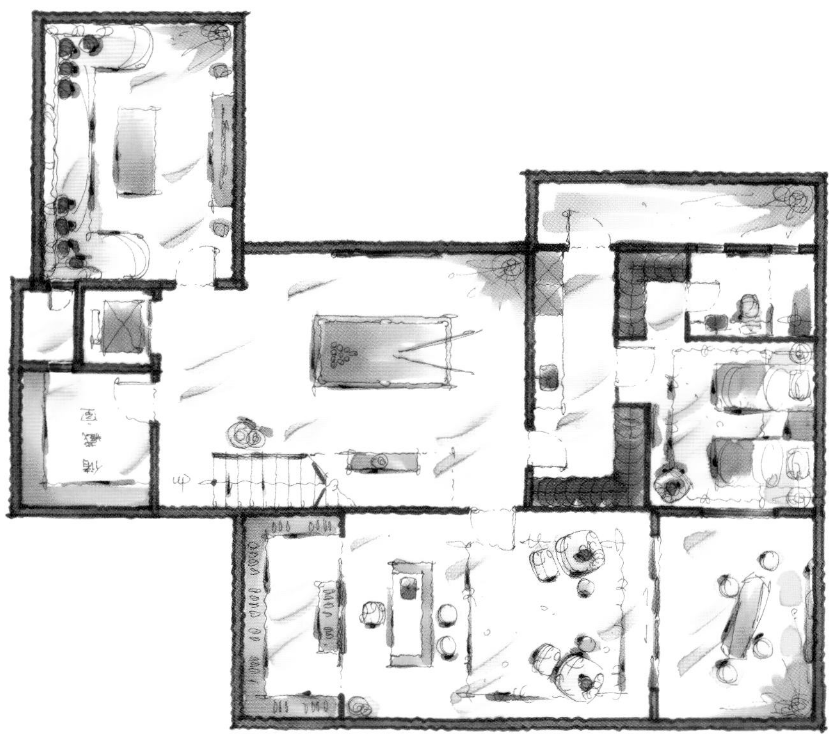

The power of design has a good influence on life. The dining room uses rationality to set up logical architecture of the space; materials such as stainless steel make the classical tone lively and fashionable; poetic sensibility enriches life; enjoyable Chinese ink painting connects ancient and modern scholars' interests of poem, wine, flower and tea. Oriental life aesthetics in poetic fragments can amaze dining time with one-thousand-year legacy and freeze the spirit and attitude.

Home is not a vanity fair and doesn't need overmuch luxuries to manifest identity. Home is also not a fashion show and doesn't need displays of sound and color under magnesium light. Home should have an aspect of amorous feeling from particular regional culture as if the bedroom of the villa which is familiar with the conciseness and rare Oriental white space habits. Lightsome and graceful, elegant and implicit, it brings modern unrestraint and nostalgic romantic temperament.

设计的力量是对生活产生好的影响，餐厅以理性建立空间的逻辑架构，不锈钢等现代材料的运用，使古典基调轻快摩登；用诗意的感性充实生活，写意的水墨画串联古今文人墨客的诗酒花茶意趣，东方生活美学在诗意化的碎片里，千年遗馈足以惊艳用餐时光，将精神与态度定格。

家不是名利场，不需要过多的奢侈来彰显身份。家也不是时尚秀场，不需要镁光灯下声色的张扬。家应该带有特定地域文化中风情流转的一面，一如别墅的卧室，深谙简约与惜墨如金的东方留白习性，轻盈婉约，清雅含蓄，带着现代的不羁与怀古浪漫气质。

Freedom is the most important part of home. Whether it is physical freedom or spiritual freedom, all can be found in the bedroom. Freedom can be as specific as color, material, furniture modeling, object size and light change in the space. Freedom can be as abstract as different feelings of four seasons from residences, casualness of life, easiness of action and deposit of thought. The daily life can be more vivid and interesting, making people release all happiness in the room.

Though senior ash forcibly occupies the designer's mind for a long time, right now it has to give place to Chinese red. Different shades of red and white combine together, which decreases seriousness and depressing atmosphere of the inhered by the closed space, brings relaxing and pleasant modern art feelings and makes the designer's artistic inspiration present splendid passion and vitality.

自由，作为家最重要的一部分，不管是身体上的自由，还是精神上的自由，都可以在卧室找到。自由可以很具体，具体到空间运用的色彩、材质、家具的造型、器物的尺寸、灯光的变化。自由也可以很抽象，抽象到居住者五观四季的不同感受，生活的随性，行动的从容，思绪的沉淀，起居坐卧都变得更生动有趣，让人在房间里释放所有的快乐。

即使高级灰霸占了设计师心头那么久，此时也不得不让位中国红，地下娱乐区，不同层次的红色与白色结合在一起，降低了封闭空间所固有的严肃感与沉闷气氛，带来轻松愉悦的现代艺术感受，让设计师的艺术灵感呈现朝霞彩霁般的激情与活力。

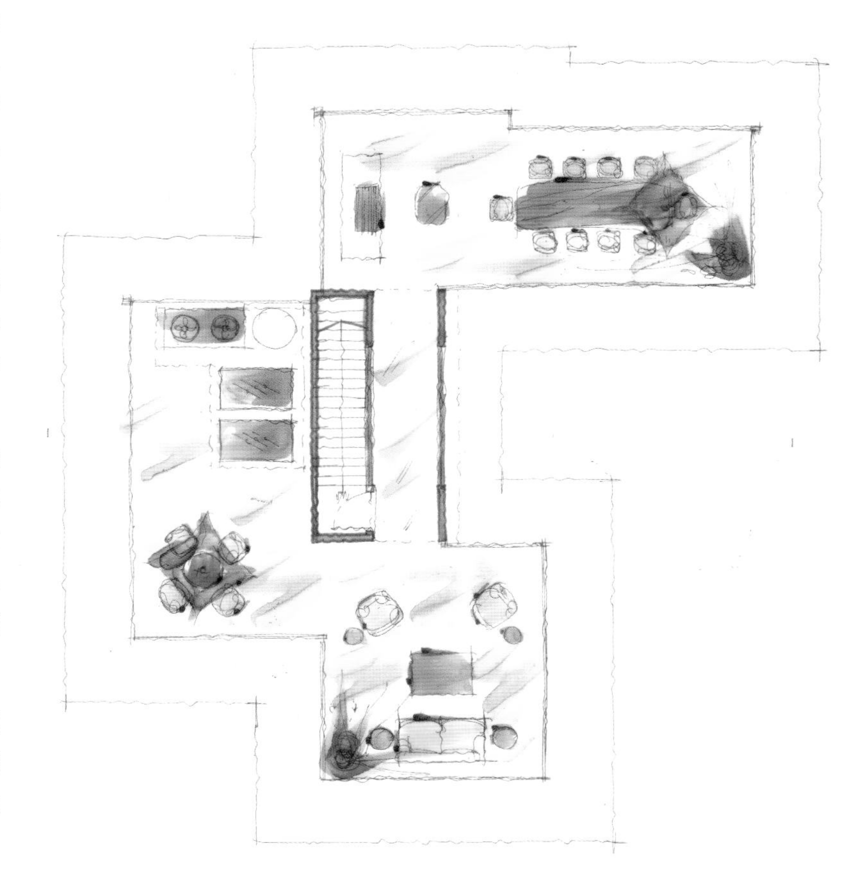

The poet Lu You wrote in *A Collection of Poems*: "Poetry is one of the six arts and should be treated seriously. If you want to learn poetry, you should work harder more than poetry and books." Zhuhai International Garden A4 Villa is beyond poetry. The most perfect decoration of home should be the owner's spirit, temperament, emotion and life interest. The most precious originality of design is to lead residents focusing more on life which moves yourselves.

陆游在《剑南诗稿》中言："诗为六艺一，岂用资狡狯？汝果欲学诗，功夫在诗外。"珠海国际花园A4别墅，诗意之外，家最完美的装饰应是主人的精神、气质、情感、生活情趣，设计最珍贵的匠心，是引居者关注感动自己的生活。

LIVING QUIETLY IN THE LANDSCAPE

幽居之时 山水之间

Design Concept | 设计理念

This is an eclectic space, using modern Chinese style to outline an inclusive space. Apart from inclusiveness, "elegance and fun" are the soul of the space. In such a space, the modern conciseness, the Chinese courtyard and the artistic conception of withered landscape present an unforgettable Zen-like beauty. In one side of the French window, the living room is bright and clean with simple color collocation, diluting the visual impact and recovering the unique clean temperament of the space. The collocation of wood color and white adds a natural flavor to the space. As the main furnishing of the space, the big blue painting becomes the regressing point of the color, the distant mountain in which fits the artistic conception of the space well.

这是一个兼容并蓄的空间，以现代中式为底，勾勒出一个包容的空间。而包容之外，"雅、趣"二字是这个空间的灵魂。这样一个空间信步走来，从现代的简洁，中式的庭院深深，到枯山水的意境，陆续呈现着让人过目不忘的禅意之美。在落地窗的一侧，客厅窗明几净，极简的颜色搭配，淡化了视觉上的冲击，却复苏了这个空间独特的干净气质。木色和白色的搭配为这个空间添上了自然的气息，而作为空间的主要装饰，一幅蓝色的大幅挂画成了空间颜色的回归点，画中的远山也很好地契合了这个空间的意境。

设计公司：福建品川装饰设计工程有限公司
主创设计师：郑陈顺
参与设计师：华淑云 、陈孝遮
项目地点：福建福州
项目面积：500㎡
摄 影 师：施凯
主要材料：大理石、挂画、鹅卵石等

As the storey height is higher, the designers extend the floor tile to the wall, making the space more open. The vertical painting, the TV background wall and the wall furnishings stress the longitudinal height of the space, making the space more open and grand. The designers bring the garden landscape into the interior, borrow the artistic conception of withered landscape and create a unique Chinese interior scenery. The pebbles are smooth as water and the stones are neat as waterside plat. The layout of the space is mainly paved with stones and pebbles of the floor. The intersection of black and white brings some casual fun, making the space reasonable. The most beautiful place is the tea area. A tree, a piece of green and a cup of tea can give you a substantial enjoyment. The Chinese framed bed is covered with simple and elegant bed sheet, near which there is a green plant, making people feel quiet. A calm heart keeps you cool. Living in such a bedroom, the restless mind can be pacified soon. Several bright color pillows, paintings, furnishings and jumping color decorations in the space avoid tediousness and present easiness.

在层高较高的情况下，设计师将地砖延续到墙面，让空间的感觉更加开阔，而竖版的挂画、电视背景墙和墙面装饰则强调了空间在纵向上的高度，让空间不减开阔大气。设计师把园林景观搬进了室内，借了枯山水的几分意境，打造出别具中国风味的室内景观。卵石如水，石材如汀，该空间的布局规划主要通过地面石材和卵石的铺设来完成，黑白交错间带着点不经意的趣味，让空间显得合理。最美的是“小汀”上的品茶区，一棵树，一片绿意，一杯好茶，就是最悠然的日子。深色木质天花板和简约的床头背景墙，给人一种很有质感的享受。中式架子床披上素雅的床单，一旁立着绿色植物，让人倍感宁静。心静自然凉，住在这样的卧室中，浮躁的心情很快得到安抚。数个亮色系抱枕、挂画、装饰品，跳跃的颜色点缀在空间中，使其规避沉闷，更显得逍遥自在。

DIALOGUE WITH VALLEY

与山谷对话

Design Concept | 设计理念

Being drunk for a thousand times in the Hibiscus curtain, having a half day off in front of the jade cliff. Contemporary elegant residences should abandon vanity and return to Oriental landscape feelings and leisure delights, with natural taste as the entire train of thought. This project intentionally obscures the definition of indoor and outdoor spaces and brings nature into Oriental comfortable space by modern method. The Chinese landscape architecture is ethereal and implicit, combining with flower, insect, bird and animal decorative elements, which eventually presents a profound Oriental cultural connotation. Designers refine traditional elements by spatial method. Whether landscape architecture or space, all present simplicity, creating a greatest interaction among nature, space and light and shadow. The style which abandons complicated details and luxury and magnificence integrates Oriental advantageous dignity and restraint, conveying rich cultural tension and inclusiveness. What's more, wood, stone and cloth present natural and tranquil temperament of the space. All are salutes to Oriental traditional culture.

项目名称：翡翠谷
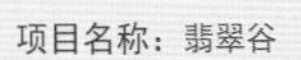
设计公司：大诺室内设计有限公司
主案设计：卢俊华
参与设计：王锦阳、曾翔、杨欣、孙晋英
项目地点：江西赣州
项目面积：480㎡
摄 影 师：邓金泉
主要材料：亚麻、饰面板、大理石等

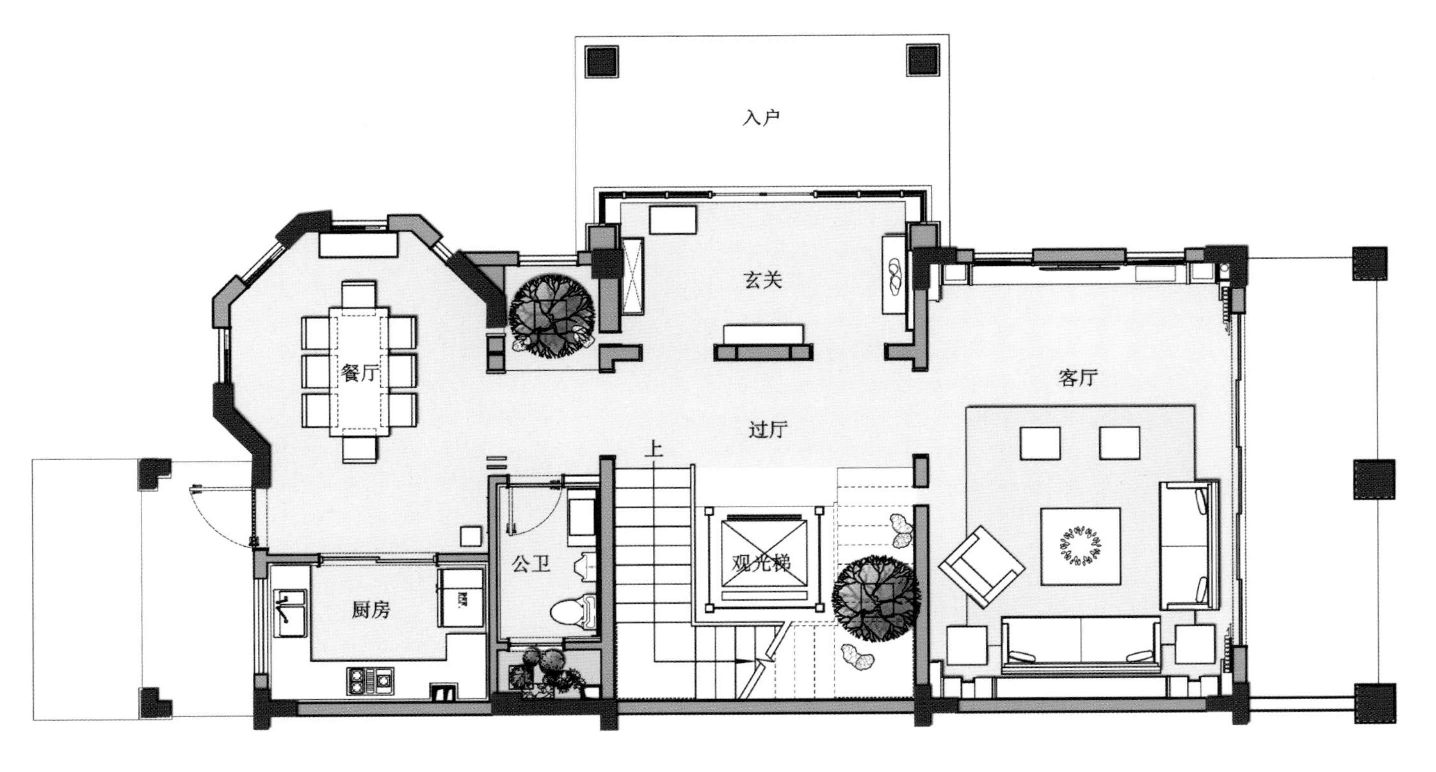
入户
玄关
餐厅
客厅
过厅
上
公卫
观光梯
厨房

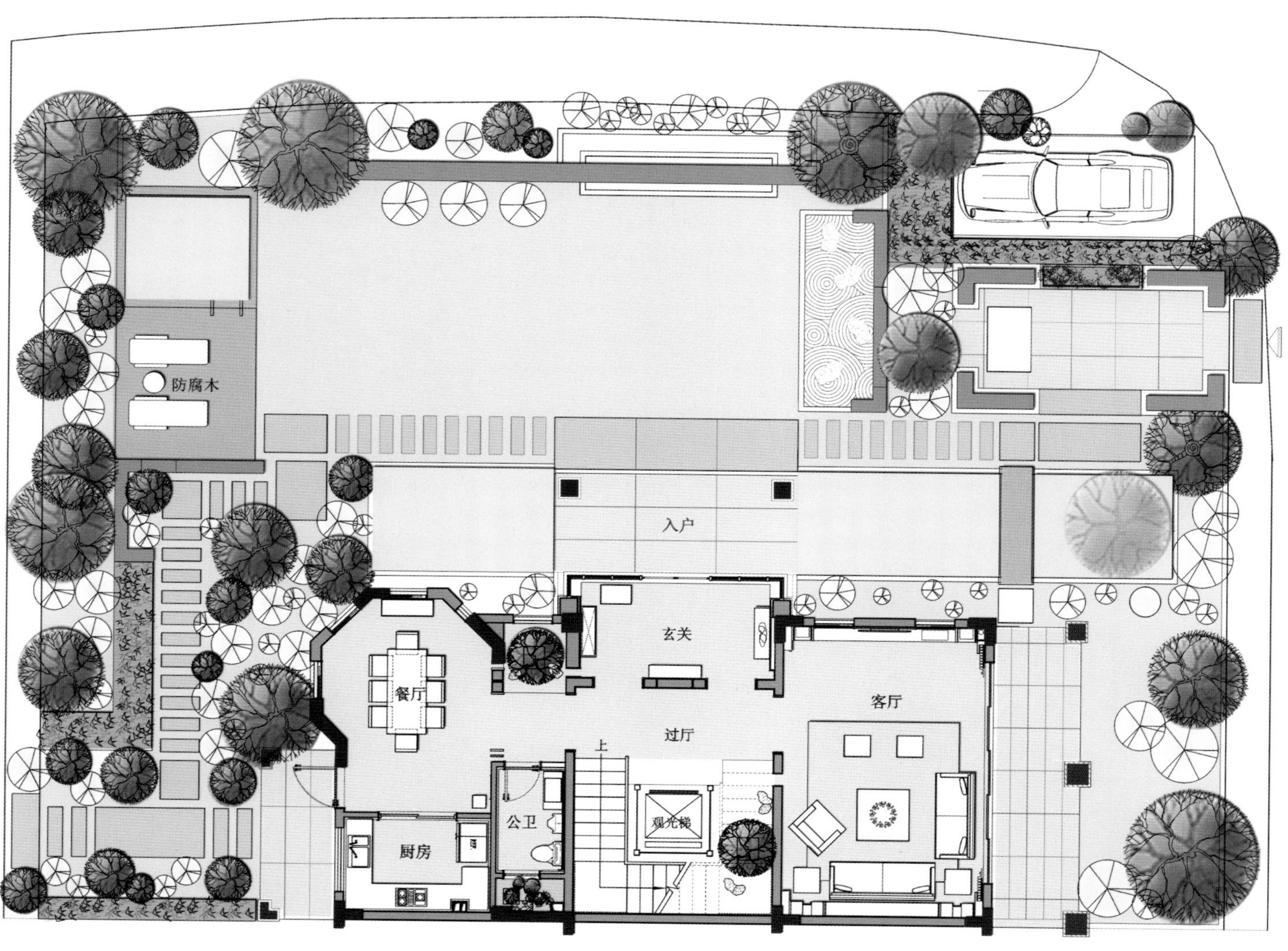

防腐木
入户
玄关
餐厅
客厅
过厅
上
公卫
厨房
观光梯

芙蓉幕里千场醉，翡翠崖前半日闲。当下雅宅应摒弃浮华，回归东方式的山水情缘，闲林野趣，以自然生息为整体思绪。本案有意模糊室内外空间界定，将自然延伸进来，施以现代手法，融入东方写意空间之中。中式造园空灵含蓄，结合花虫鸟兽元素装饰，最终呈现出一份深入骨髓的东方文化内涵。设计师将传统元素以空间手法进行提炼，无论是造园，还是空间，都体现简约，让自然、空间、光影形成最大程度上的互动。摒弃繁复细节和奢华雍容的风格，融入东方得天独厚的庄重内敛，兼收并蓄，传递丰富的文化张力与包容性，并用木石布体现空间自然宁静的气质，是本案对东方传统文化的致敬。

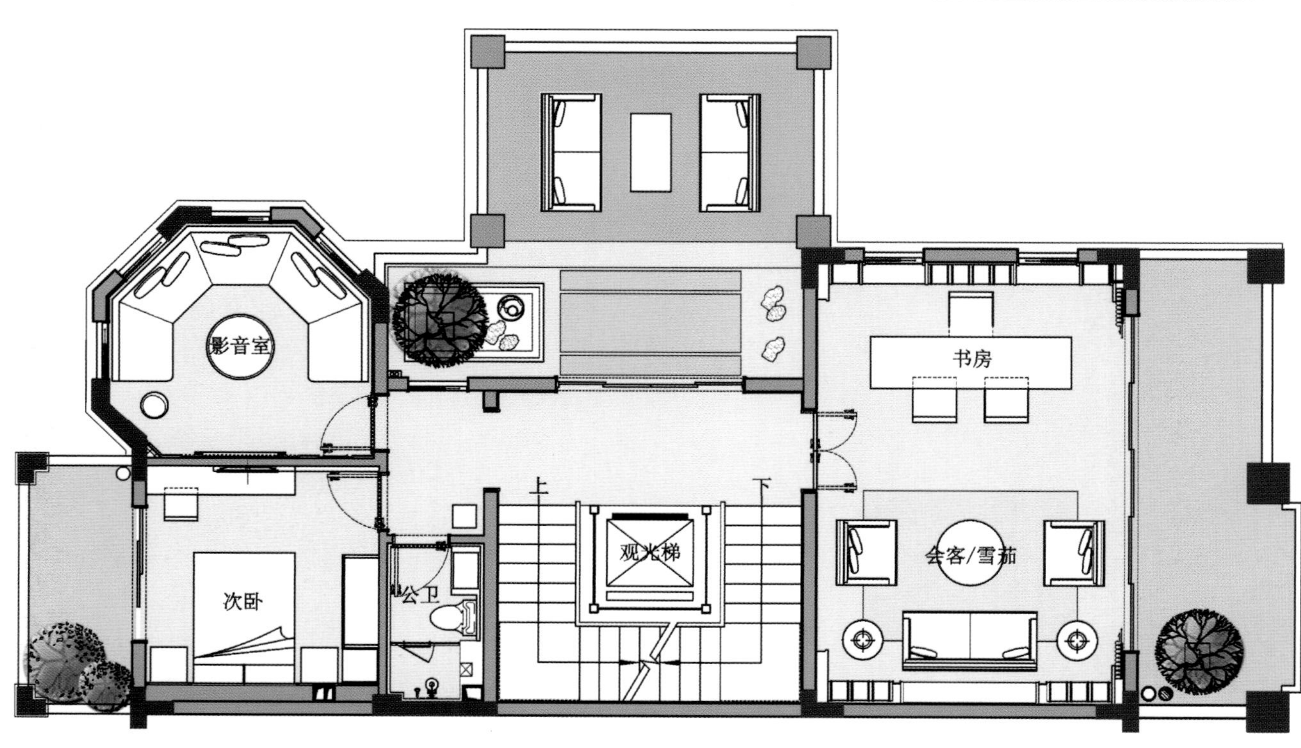

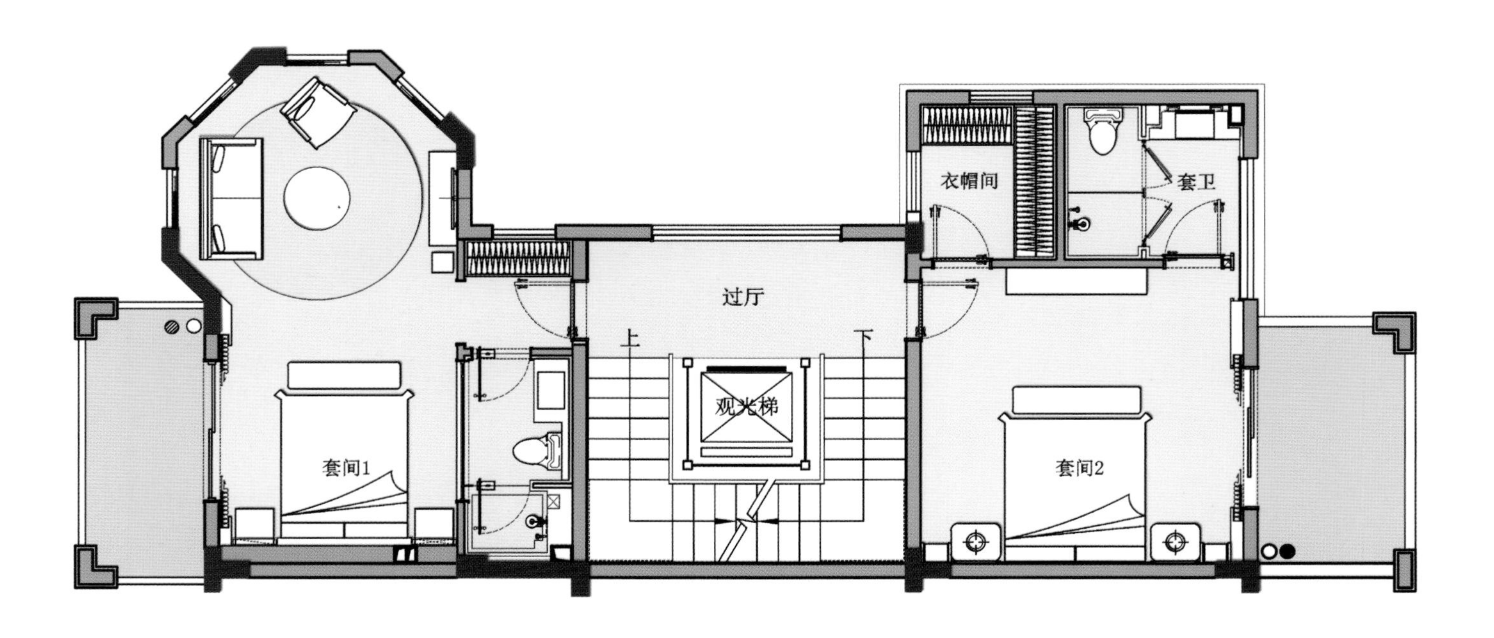

衣帽间
套卫
过厅
上
下
观光梯
套间1
套间2

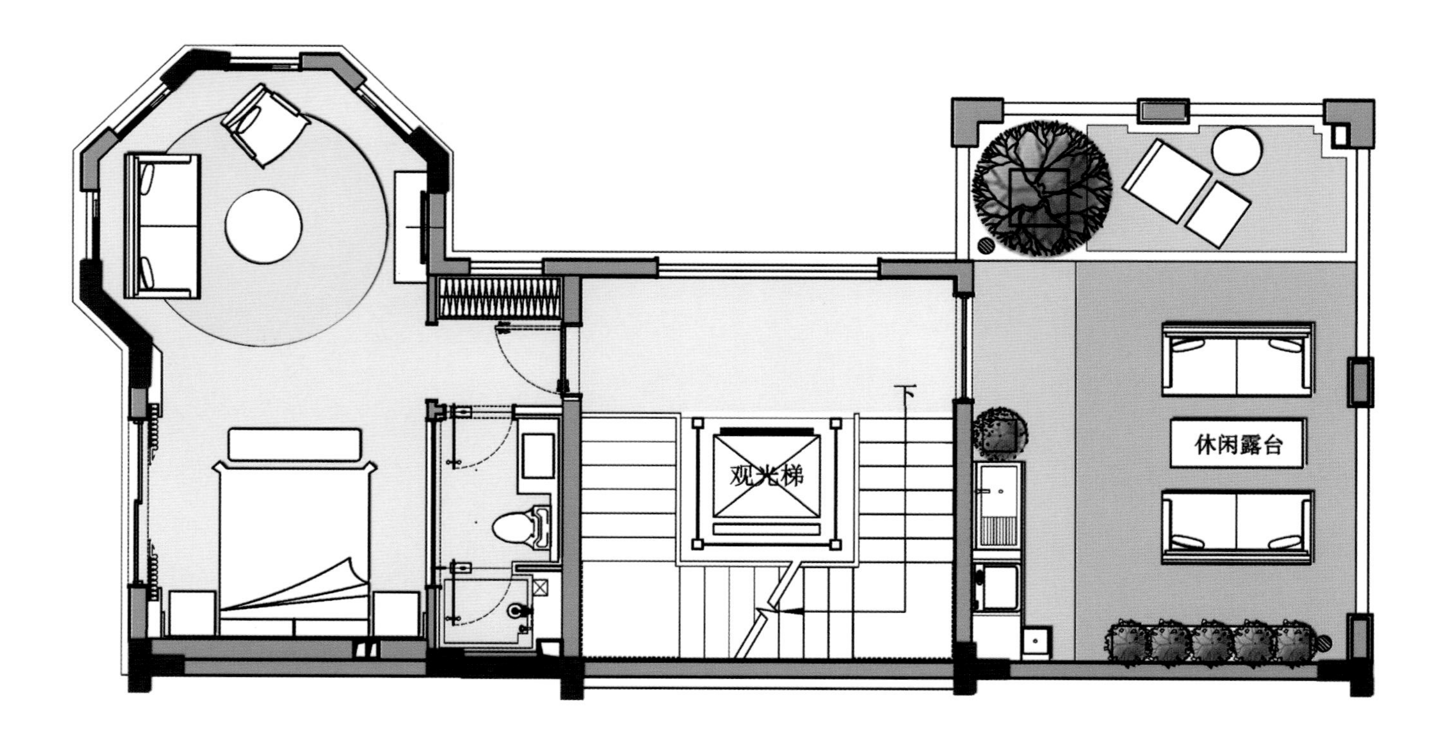

观光梯
休闲露台
下

MODER

N STYLE

现代风格

LUXURIOUS AND ELEGANT STYLE, MAGNIFICENT MANNER

奢雅风范 恢弘气度

Design Concept | 设计理念

The designer firstly needs to carry out the functional regions and reasonable renovation of this building of nearly two thousands square meters to increase the plasticity and practicability of the space and to reflect the design art expression.

The first impression of the villa is magnificent, elegant, luxurious and spacious. The ground and wall of the first floor are covered with Turkish rose marbles, setting a low-key and luxurious tone for the space. Elegant spiral stairs mainly use wire-wove marbles and combine with cambered glass and leather stitching armrest, manifesting the exquisite manner. The designer dismantles and regroups the space and retains six columns of the original architecture, making the space spacious, comfortable and magnificent. Luxurious crystal droplight makes the space shine. In the chatting and rest areas at two sides, customized furniture and handmade carpet add beauty to each other.

项目名称：益田长春喜来登会所别墅
设计公司：DHA香港洪德成设计有限公司
设 计 师：DHA香港洪德成设计团队
项目地点：吉林长春
项目面积：1980㎡
摄 影 师：王辉
主要材料：黑郁金香大理石、泰姬灰大理石、玉森林大理石、金丝白玉大理石、文化石、紫檀木地板、虎檀尼斯木饰面、山纹橡木饰面等

设计师首先需要对将近2千平方米的建筑内部空间进行功能分区及合理改造，增强空间的可塑性和实用性，同时以便体现设计艺术的表达。

走进别墅的第一观感是：大气、典雅、奢阔。一层地面和墙面运用土耳其玫瑰大理石，奠定空间奢华基调。优雅的旋转楼梯主要使用金丝网纹大理石，结合弧形玻璃和皮革车线扶手，精品气度一览无余。设计师在重新拆合重组空间时，保留了原建筑的6根石柱，令空间宽敞舒适的同时更显恢弘。而豪华的水晶吊灯则使整个空间神采奕奕。两侧的洽谈休息区，定制家具与手工地毯相得益彰。

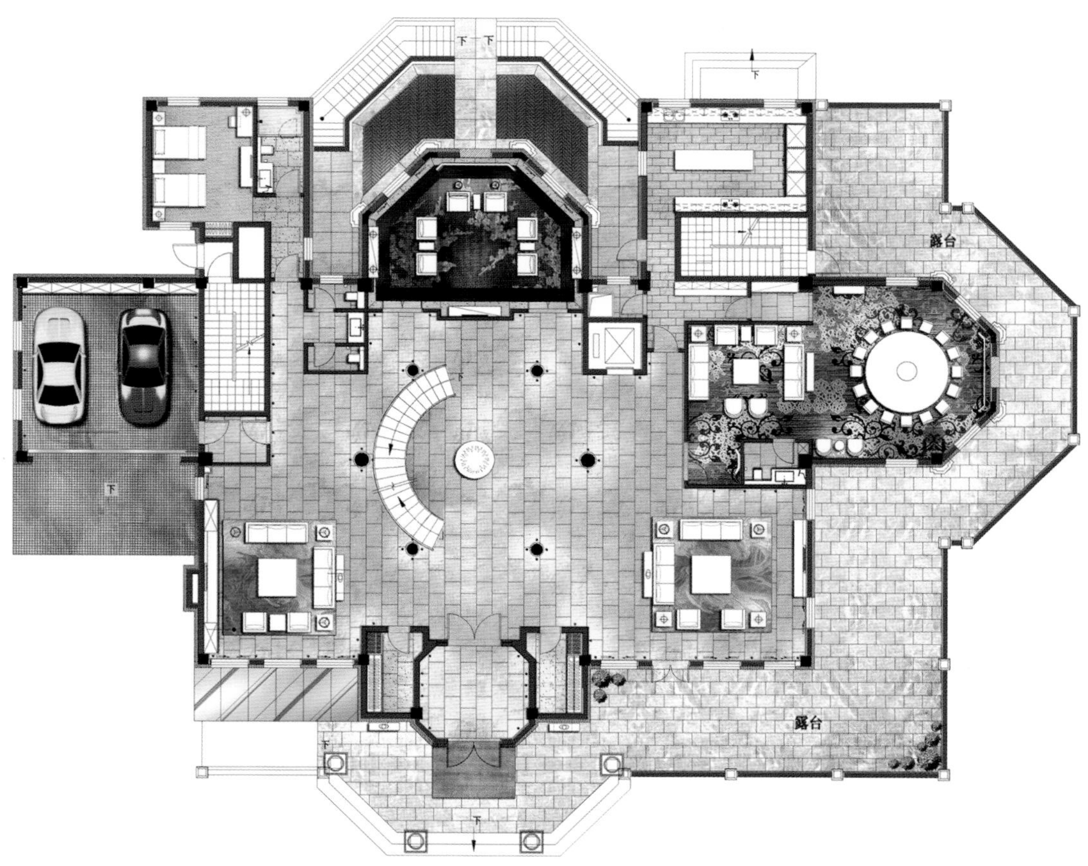

露台
露台

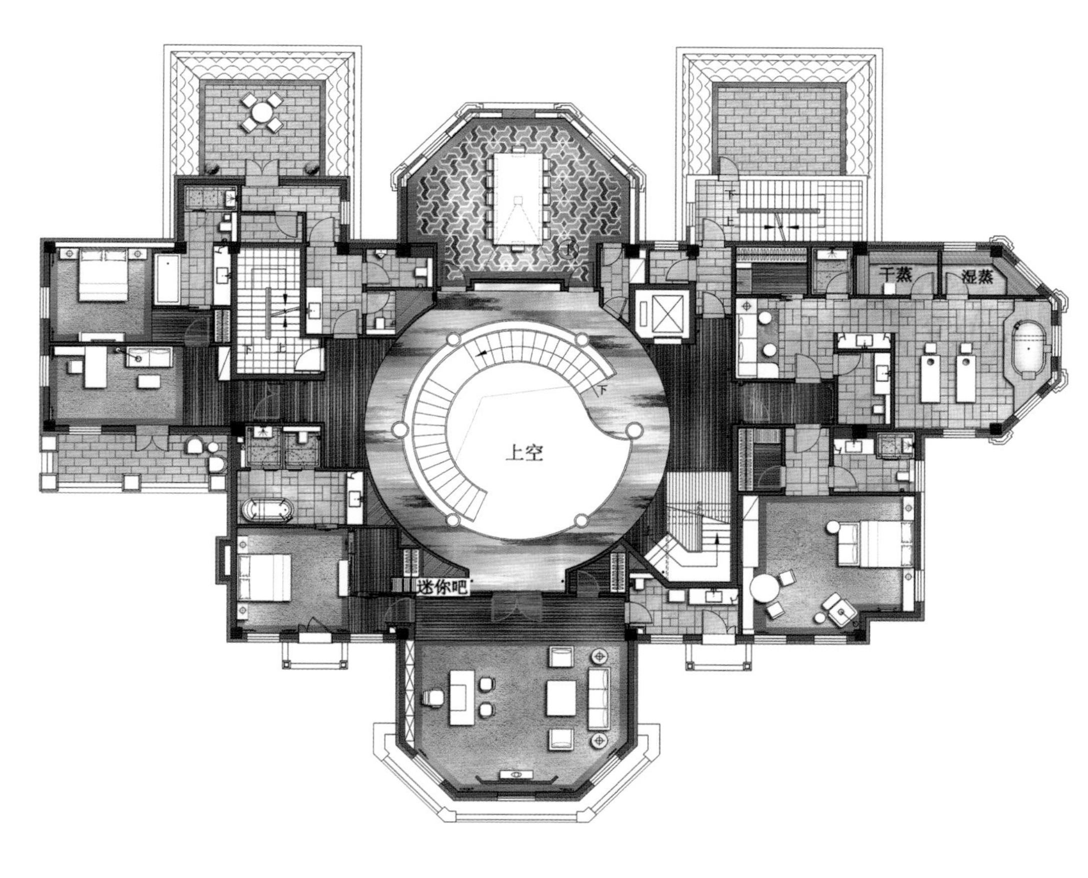
干蒸
湿蒸
上空
迷你吧

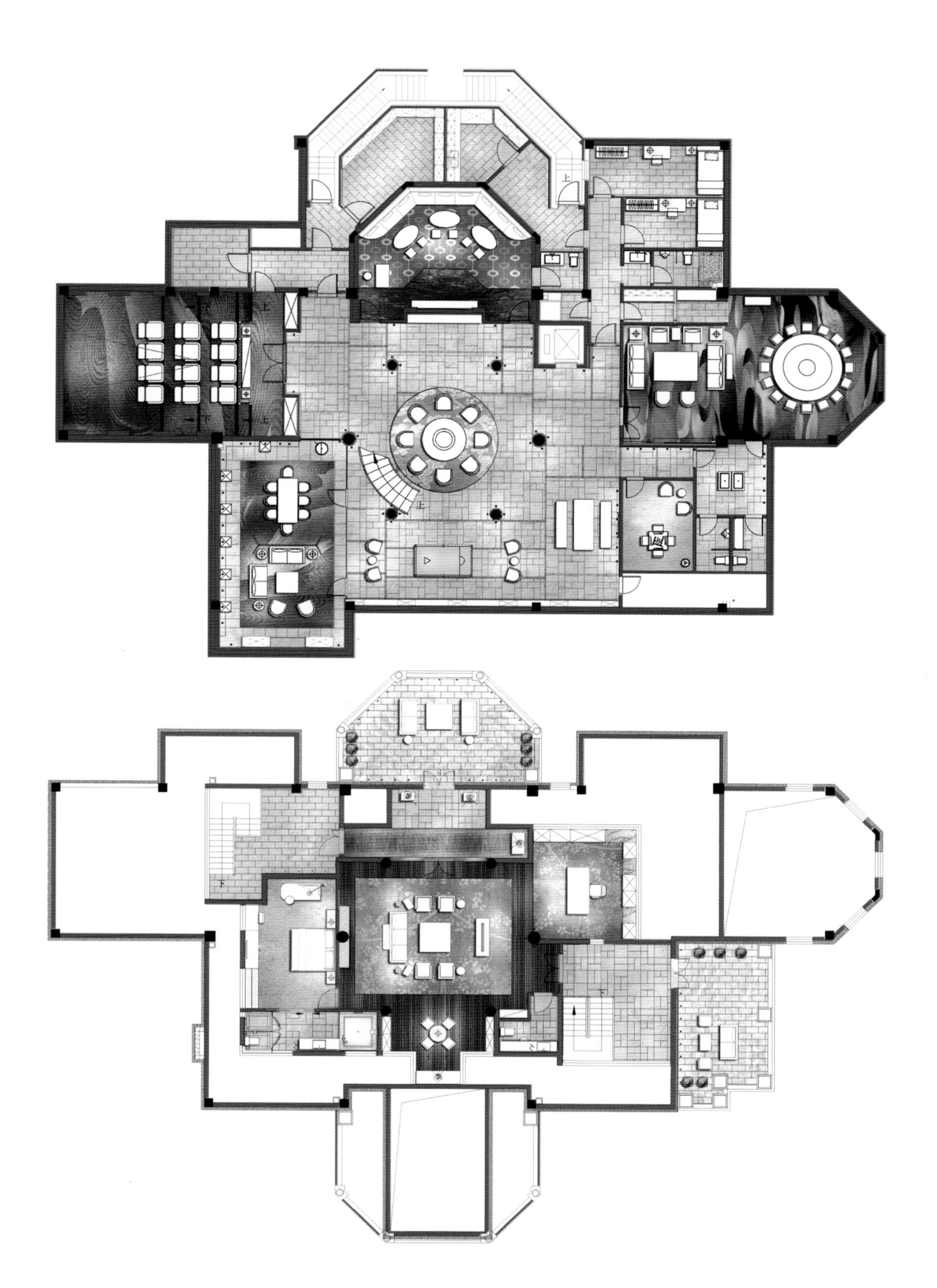

The negative layer is mainly for entertainment, leisure and dining. Elegant leisure bar, red wine cigar display area, private cinema, chess and card room, dining room and Chinese tea room, extremely enjoyment is everywhere. In particular the Chinese tea table is constituted by two even parts of the whole tree with a natural temperament and manner. The second floor is the extension of the first floor reception function along with accommodation function. The penthouse top space is relatively independent. The designer flexibly uses European leisure and comfortable sense and natural and decent quality.

Throughout the whole villa, the designer considers and carefully deliberates designs of every floor, making perfections more perfect.

负一层以娱乐休闲餐饮功能为主。优雅的休闲吧、红酒雪茄展示区、私人影院、棋牌室、餐饮房和中式品茶区……极致享受无处不在。特别值得一提的是中式茶木桌，以整棵树凿开两半拼制而成，自有种天然的气质和风范。二层则是一层空间接待功能的延伸，并具备住宿功能。阁楼顶层空间相对独立，设计师灵活采用了欧陆式的休闲舒适感和自然大气的品质感。

纵观全墅，每一层的设计都经过设计师长时间的思考和细细推敲，可谓精益求精。

DREAMLIKE POETRY

梦幻的诗意

Design Concept | 设计理念

This project skillfully integrates Chinese elements with modern materials and expresses the yearning and pursuit of dignified and elegant Oriental spiritual state through the inheritance and integration of quintessence in Chinese style. The application of ink grain marble, metal and wood veneer and the combination of Zen flower arrangement, potting and furnishing article couple hardness with softness and are innovative and changeable, supplemented by landscape objects, which is flexible and full of dreamlike poetry. The unified gold tone in rich texture and decorations full of artistic conceptions endow the space with unique Oriental charm. The living room presents clean and neat sense of line; the sofa has a familiar antique charm. The tearoom uses pine bonsai, tea sets with ink painting and a large landscape painting to create quaint and quiet tea atmosphere. The droplight of the staircase through upstairs and downstairs strengthens the exchange of spaces.

本案将中式元素与现代材质巧妙兼揉，通过中式风格精髓的传承与融合，借以表达对端庄丰华东方式精神境界的向往和追求。空间中水墨纹大理石、金属、木材饰面的运用，禅意插花盆栽摆件的组合，刚柔并济，创新赋予变化，辅以山水物象融入空间，灵动而富有梦幻的诗意。空间统一的金色调富有质感，充满意境的装饰让别墅流动着别样的东方韵味。客厅呈现出明净的线条感，沙发有着似曾相识的古韵，茶室以青松盆景、带有水墨画的茶具和大幅山水画营造出古雅安静的品茗环境，楼梯间的吊灯贯穿楼上楼下整个区域，加强了空间的交流。

项目名称：金地集团湖山花园别墅
设计公司：深圳市大集空间设计有限公司
主创设计：巢宇、曾广辉
参与设计：李坤华、练腾元
项目地点：广东东莞
项目面积：231㎡
摄 影 师：刘虎
主要材料：浅色橡木木地板、影木木饰面、玫瑰钛不锈钢、波斯灰大理石、水云　沙大理石等

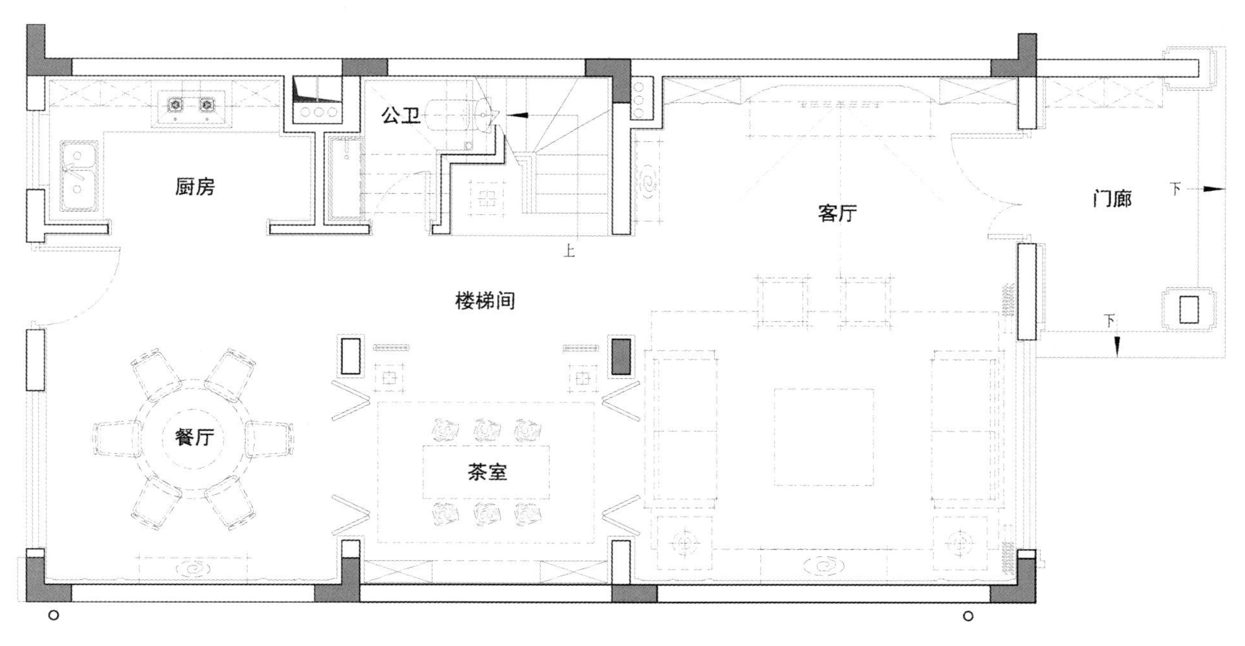

公卫
厨房
客厅
门廊
下
上
楼梯间
下
餐厅
茶室

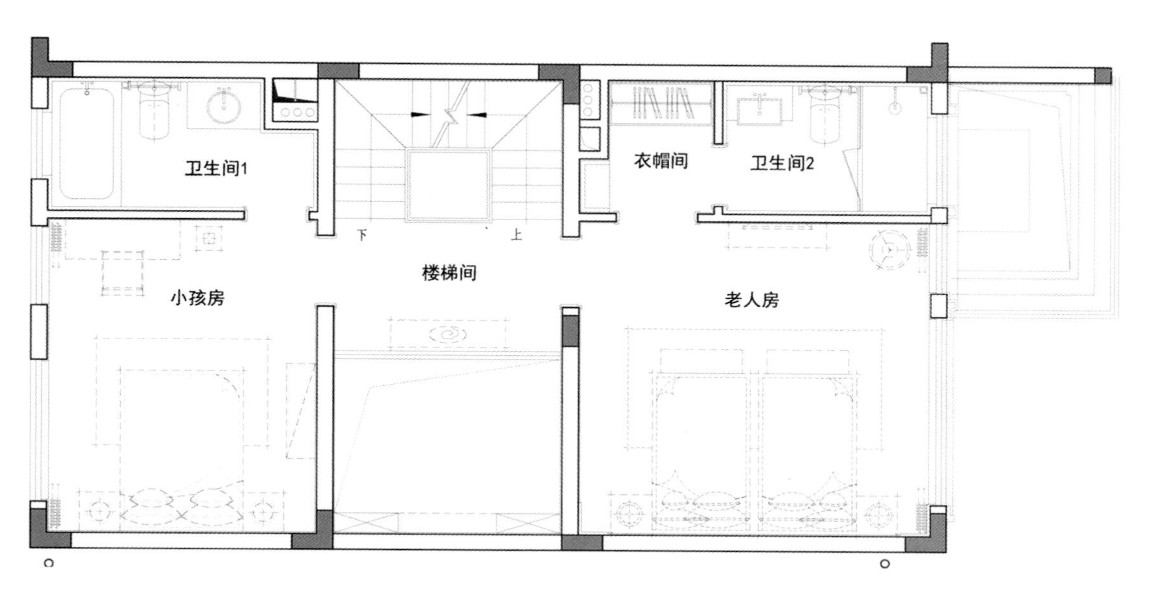
卫生间1
楼梯间
衣帽间
卫生间2
下
上
小孩房
老人房

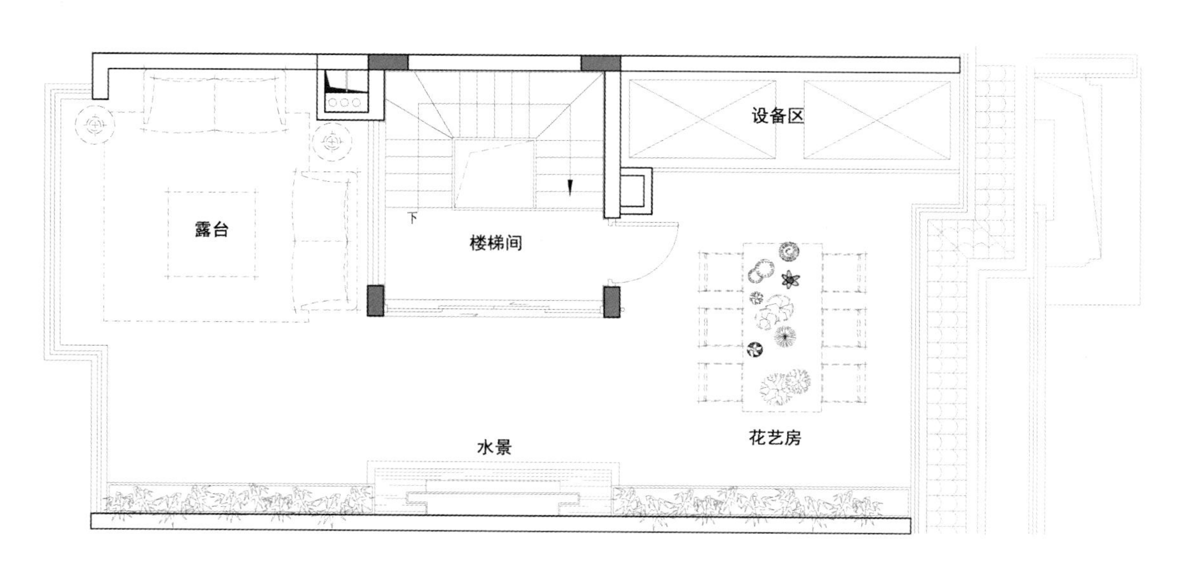
设备区
露台
下
楼梯间
水景
花艺房

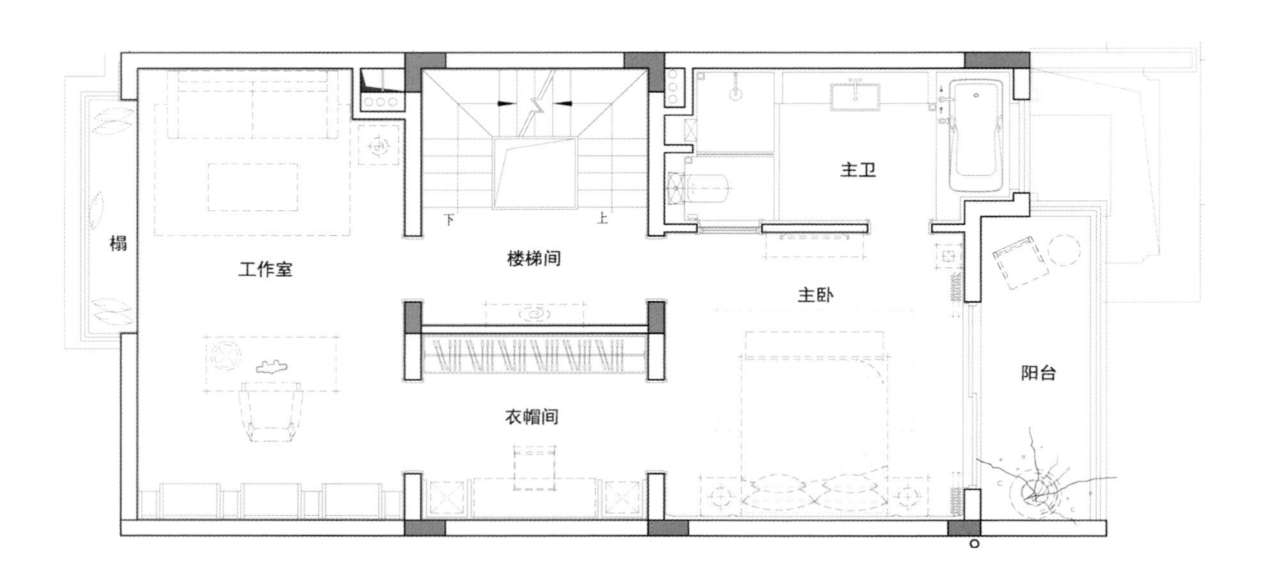
榻
工作室
楼梯间
下
上
主卫
主卧
阳台
衣帽间

DANCING SILVER LEAF, LEISURE TIME

银叶飞舞 慵懒时光

Design Concept | 设计理念

The two-story high living room uses droplight to fill the empty sense of the space with appropriate white space to keep some imaginations. This is leisure life with flowers and sunshine; the once desperately pursued world is in front of your eyes now. The main part is natural transparent stone with sense of mystery which contains extremely fine craftsmanship of nature. Light sensation is presented through stone, like a fairyland, as if flying in the top of the clouds while the opposite is a piece of dancing silver leaf; fresh contrast brings new feelings and creates comfortable and tranquil life. Conciseness achieves aesthetics. The designer uses lines to create rigorous, right and meticulous designs, manifesting fashionable and magnificent luxurious style. We choose wood veneer, wood floor, upholstery and mirror steel to promote the texture of the space; thin lines create different senses of layering and present fashionable and delicate life taste.

两层挑空的客厅用华丽的垂直吊灯来填充空间的空旷之感，但又适当地留白，保留一点想象空间。这里充斥着鲜花、阳光、慵懒的生活，曾几何时人们苦心追寻的世界如今就在眼前。主幅是天然的透光石材，蕴藏着大自然鬼斧神工的神秘感，光感透过石材展现出来，羽化登仙，似乎凌驾在云彩的尖端，对面却是一片银叶飞舞，鲜明的对比带来新的感受，营造出舒适恬淡的生活意境。简约成就美感，用线条构造严谨、端正、一丝不苟的设计，彰显着时尚大气的的奢华风范，我们选用木饰面、木地板、扪皮、镜钢提高空间的质感，细细的线条拉出不同的层次感，展现了既时尚又细致的生活品位。

项目名称：依云雍景湾别墅样板房
设计公司：广州源创榀格环境艺术设计有限公司
设 计 师：周文胜
项目地点：广东佛山
项目面积：844㎡
摄 影 师：戴胜念、梁锡坚
主要材料：石材、墙纸、扪皮、木饰面、镜钢、黑镜等

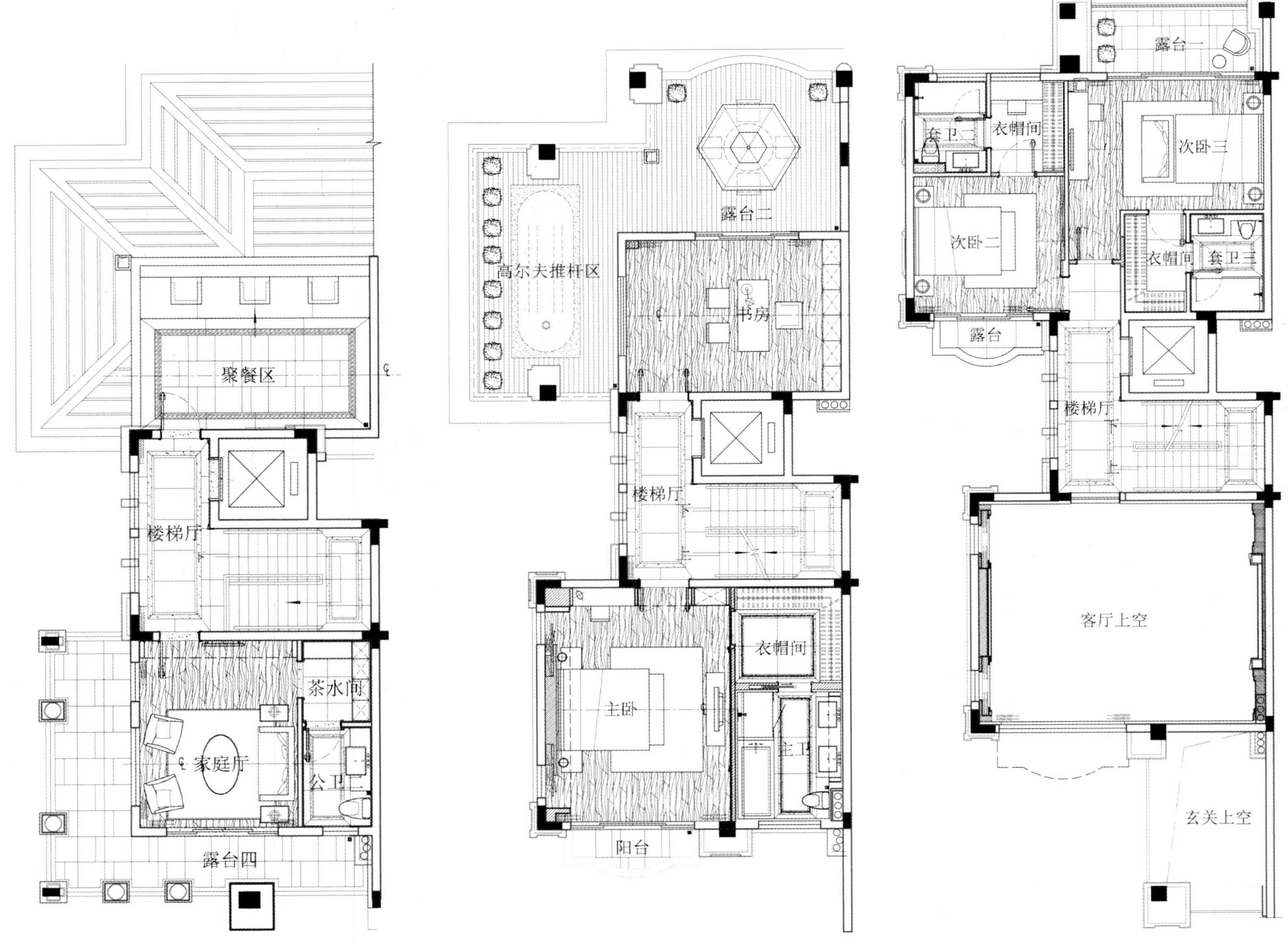

聚餐区
楼梯厅
茶水间
家庭厅
公卫
露台四
露台二
高尔夫推杆区
书房
楼梯厅
衣帽间
主卧
主卫
阳台
露台一
套卫二
衣帽间
次卧三
次卧二
衣帽间
套卫三
露台
楼梯厅
客厅上空
玄关上空

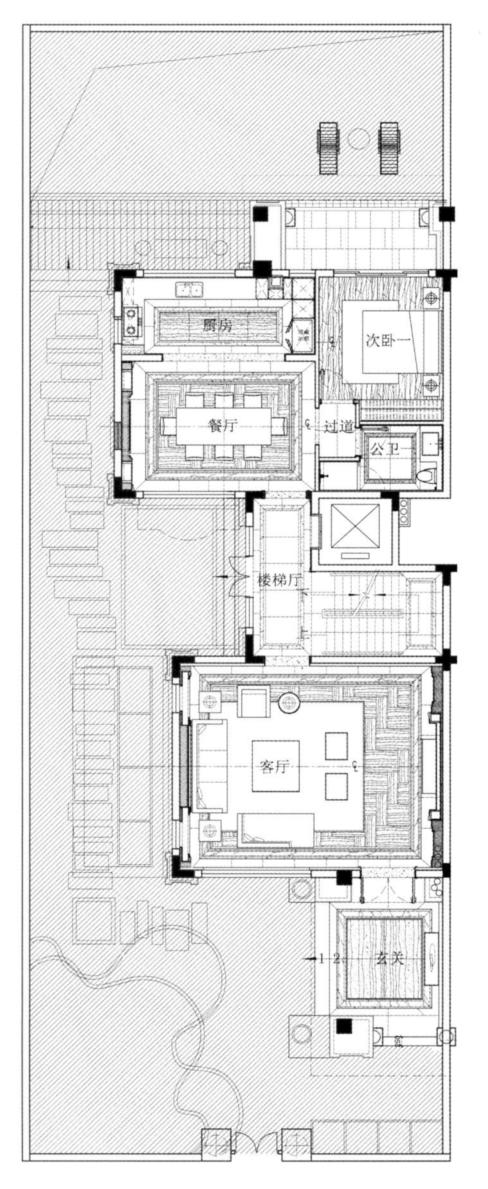

厨房
次卧一
餐厅
过道
公卫
楼梯厅
客厅
玄关

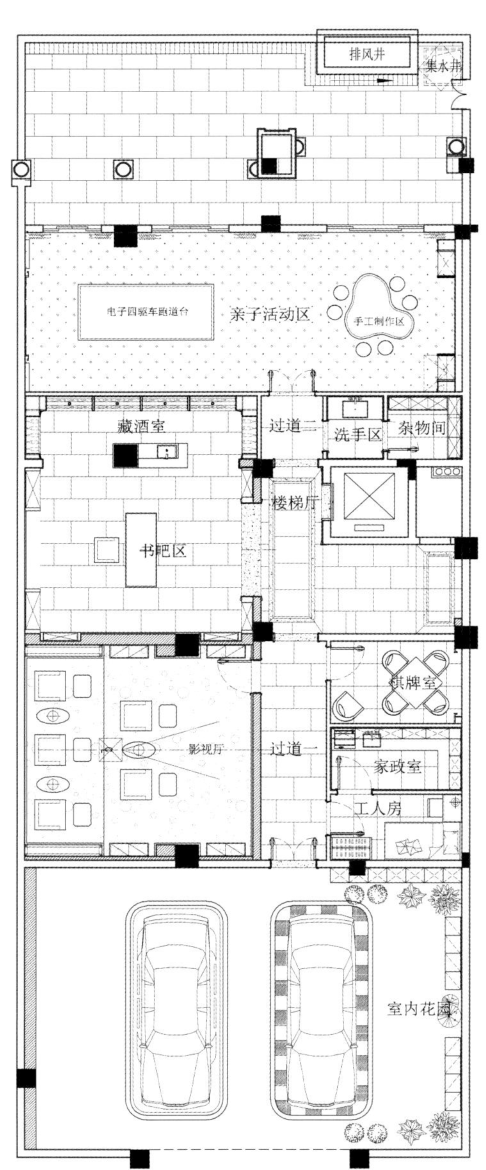

排风井
集水井
电子四驱车跑道台
亲子活动区
手工制作区
藏酒室
过道二
洗手区
杂物间
楼梯厅
书吧区
棋牌室
影视厅
过道一
家政室
工人房
室内花园

12345678910

EXPERIENCING PROSPERITY, EMBRACING LIFE

阅尽繁华 拥抱生活

Design Concept | 设计理念

This villa is located is a cultural tourism city in the outskirts of Beijing. The style of the project is Hong Kong neo-classical style. It inherits traditional elements at the same time evolves "golden age", interprets contemporary elite class's master of honors and life state of getting away from the the Madding Crowd and embracing happiness.

Opening up the door, the porch with artistic flavor finds you, as if opening the prelude of living experience. New baroque furniture pattern matches with oil paintings of Vincent van Gogh; the delicate gesture manifests gorgeous aesthetics and injects art temperament into life.

别墅位于北京郊区一个文化旅游城，项目风格为港式新古典，在继承传统元素的同时演化出"黄金时代"，诠释当代精英阶层驾驭荣誉的同时远离凡尘喧嚣，拥抱幸福的生活姿态。

推开大门，极具艺术气息的玄关扑面而来，仿佛拉开了入住体验的序幕。新巴洛克风格的家具款式，配以梵高的油画作品，细腻姿态，彰显华丽美感，把艺术气质注入生活。

项目名称： 北京恒大文化旅游城B4-6别墅
设计公司： 百搭园软装
项目地点： 北京
主要材料： 不锈钢、木地板、壁纸等

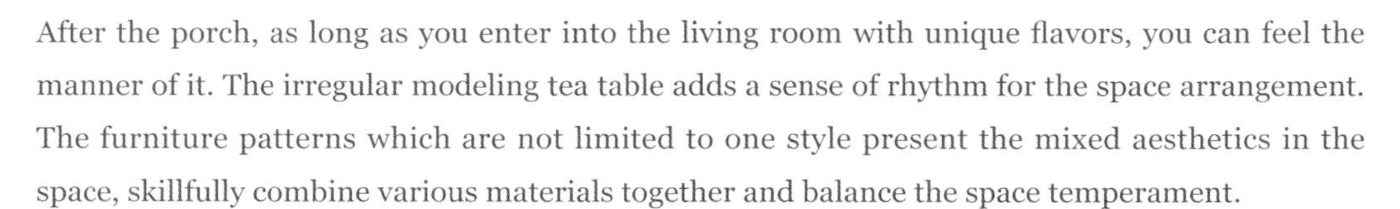

After the porch, as long as you enter into the living room with unique flavors, you can feel the manner of it. The irregular modeling tea table adds a sense of rhythm for the space arrangement. The furniture patterns which are not limited to one style present the mixed aesthetics in the space, skillfully combine various materials together and balance the space temperament.

When night falls, the family sit around the sofa and laugh without concerning the outside trivial matters. The designer integrates royal blue and orange into the space as if the spacious feelings with which the space culture contains amorous feelings in different places, presenting dreamy colors of modern city. Modern sofa matches with art corner table with neo-classical flavors, promoting the owner's height of pursuing exquisite life. Distinctive small artworks under the tea table foil the life tonality.

经过玄关，迈入韵味独特的客厅，立即就能感受到客厅彰显的气度，不规则造型的茶几增添空间布置的律动感，不拘泥一种风格的家具款式，在空间里展现混搭美学，将多种材质精妙的组合在一起，既有独立的美感，也相互平衡了空间气质。

夜幕降临，一家人围着在沙发上谈笑风生，外界一天的繁杂远抛云霄。设计师将皇室蓝、橘黄以及妃红融入空间中，宛若空间文化包容异地风情的宽阔情坏，展现出现代都市的梦幻色彩。用现代感十足的沙发搭配新古典韵味的艺术角几，提升了主人对精致生活追求的高度。茶几上造型独特的小艺术品烘托生活格调。

In the dining room, a "tasty" orange flower triggers people's appetite instantly. Modern geometric irregular white droplight matches with dazzling gold tableware furnishings, making the dining room shine luxurious, modern and intoxicating charms of international urban. Arc lines with classical beauty of the dining chairs soften the hale straight lines of the dining table. The collision of yellow and blue adds dining atmosphere for the space. Classical art paintings on the wall bring your thoughts back to ancient time between the looking back and laughter.

After dinner, you can habitually go to the study to lean on the table and conveniently open up a favorite book, thus the emotions are surging. The horse head artwork and oil paintings by Vincent van Gogh promote the space temperament. Collecting favorite books and art works is a kind of fun. Pen, ink and book, all is well.

来到餐厅，一抹“开味”的橘黄色花艺，让人瞬间食欲大增。现代几何造型的不规则白色吊灯搭配金色耀眼的餐具配饰，让餐厅空间闪耀着国际都会奢华摩登的醉人魔力。餐椅古典之美的弧形线条，柔和了餐桌硬朗的直线。黄与蓝的对撞增添了空间的用餐氛围，墙上古典艺术挂画在回眸谈笑之间，令思绪不经意穿越到远古时代。

晚餐过后，习惯性迈入书房沉静一会儿倚坐在桌前，随手翻开一本喜欢的书籍，不禁思绪涌动……马头的艺术品和梵高的油画提升了空间气质。收藏喜爱的书籍、艺术品也是一种乐趣，笔墨书卷，浓妆淡抹总相宜。

The comfortable and quiet master bedroom creates vision on the basis of modern aesthetic pursuits. The details are not lack of representative style presentation. It abandons trivial matters and noise, floating, tranquil and beautiful, profound yet not meticulous. Casually lying in the comfortable bad and looking up to the sky through the window, you can think a lot.

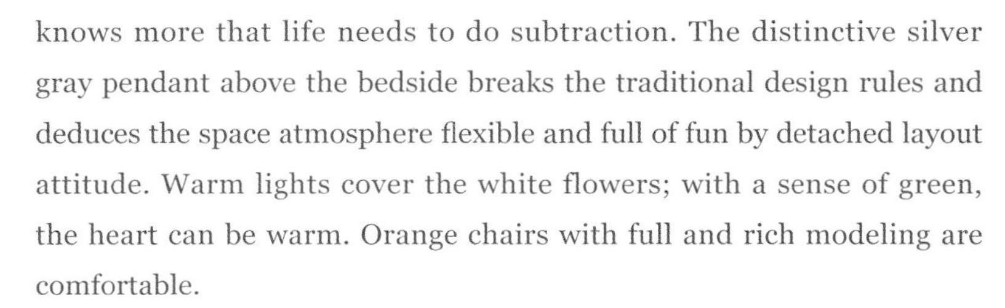

Moving to the elder's room with security sense, the experienced life knows more that life needs to do subtraction. The distinctive silver gray pendant above the bedside breaks the traditional design rules and deduces the space atmosphere flexible and full of fun by detached layout attitude. Warm lights cover the white flowers; with a sense of green, the heart can be warm. Orange chairs with full and rich modeling are comfortable.

进入舒适静谧的主卧，以现代审美诉求为标准进行视觉营造，细节处当然不乏代表性的风格呈现，去除繁杂与喧嚣，浮动却又静美，博大却不乏精细。随性躺卧在舒服的大床上，透过落地窗仰望星空，浮想联翩。

挪步到充满安定感的长辈房，有阅历的人生更懂得生活需要做减法。床头上方极具个性的银灰色挂饰，设计突破传统的规矩，以超然物外的格局态度，将空间的氛围演绎得灵活而富有意趣。暖色的灯光笼罩着白色的花朵，点缀一抹绿意，心里也暖暖的。橘黄色座椅，造型饱满丰盈，舒适感极好。

The youth's room reveals the personality, the youth and the passionate time. Geometric surface of the bedside cabinet with characteristics matches with linear stereo table lamp; for the youth who is in the age of flowering season, it is definitely a display of self-personality.

然后是青春年华，激情岁月，彰显个性的青年房，床头柜的几何切面，个性十足，配上线性立体造型的台灯，对于正值花季年龄的青年来说无疑是一种自我个性的展示。

AME
STY

RICAN

LE

美式风格

TRANQUIL AND PEACEFUL TIME

时光静好 岁月无争

Design Concept | 设计理念

Accustomed to the flashy downtown, one always longs a little peace. The designer always shoulders this kind of mission to create a variety of different spaces and to percept human changes and the existence of graceful life. Here the designer is the carrier of nature; this kind of carrying is the imitation and migration of nature essences.

Respect the original state of life, change its form and not change its attribute. The designer quietly brings natural elements into interior to create a scene and tell a story in a specific space. The ceiling of the living room is log groove modeling; the walls are original stone splicing. The designer retains the primitive form of materials on the basis of modern style, making people feel in nature when in interior.

看惯了浮华闹市，总渴望一丝宁静。设计师一直都肩负着这样的使命，打造各种不同的空间，感知人间冷暖，生命曼妙之所在。在这里，设计师就是大自然的搬运工，而这种搬运，是自然精髓的仿造与迁徙。

尊重生命的原始状态，改变其形，不变其性。设计师将自然元素悄然无声引入室内，在一个特定的空间里去塑造一个情景，讲述一个故事。客厅天花为原木材料凹槽造型，墙面造型为原始石材造型的拼接，设计师力求在现代风格的基础上保留材质最原始的形态，让人身在室内，也能有着亲临大自然的心境。

项目名称：湘府世纪39#H-3户型别墅示范单位
设计公司：本则创意（柏舍励创专属设计）
项目地点：湖南长沙
项目面积：480m²
主要材料：木饰面、仿石砖、皮革、大理石等

The designer skillfully divides every function space to endow every space with unique elements. Sun lights sparkle over the well-organized background wall of the dining room, creating mottled light and shadow. The comfort sense of log makes people feel pleasant. Design of the kitchen is humanized; larger activity space and bright lights provide an enjoyment before dining time.

设计师巧妙地划分各个功能空间，让每个空间都有独特的元素呈现。餐厅处高低错落的背景墙，透过层层阳光，斑驳的光影，原木的舒适感让人的心情不由随之愉悦。厨房的设计也是极为人性化，较大的活动空间和明亮的光线，餐前时光也是一种享受。

Styles of bedrooms are different with relatively warm tones. The collocation of leather and leopard prints reveals light Euro-American flavor, tranquil and wild, which is perhaps the integration of modern design concept and Euro-American style by the designer. The entire study is elegant; all decorations such as wheat color wallpaper and solid wood book shelf are original. Reading a good book and listening to a melody, this is the tonality of life. In this materialistic era, people not only pursue a functional space to live, what's more, it can carry emotions and express a life attitude.

卧室的风格不尽相同，色调较为温馨，皮革和豹纹的搭配透露淡淡的欧美风情，静谧中夹着一丝野性，也许这也是设计师对于现代设计理念与欧美风格融合之处。书房较整体而言淡雅许多，麦色的墙纸，实木书架，一切装饰都是原汁原味，品读一本好书，倾听一段旋律，这便是生活的格调。在这个物欲充裕的年代，人们追求的不仅仅是一个可以栖身的功能性场所，更重要的是，它寄托情感，表达一种生活的态度。

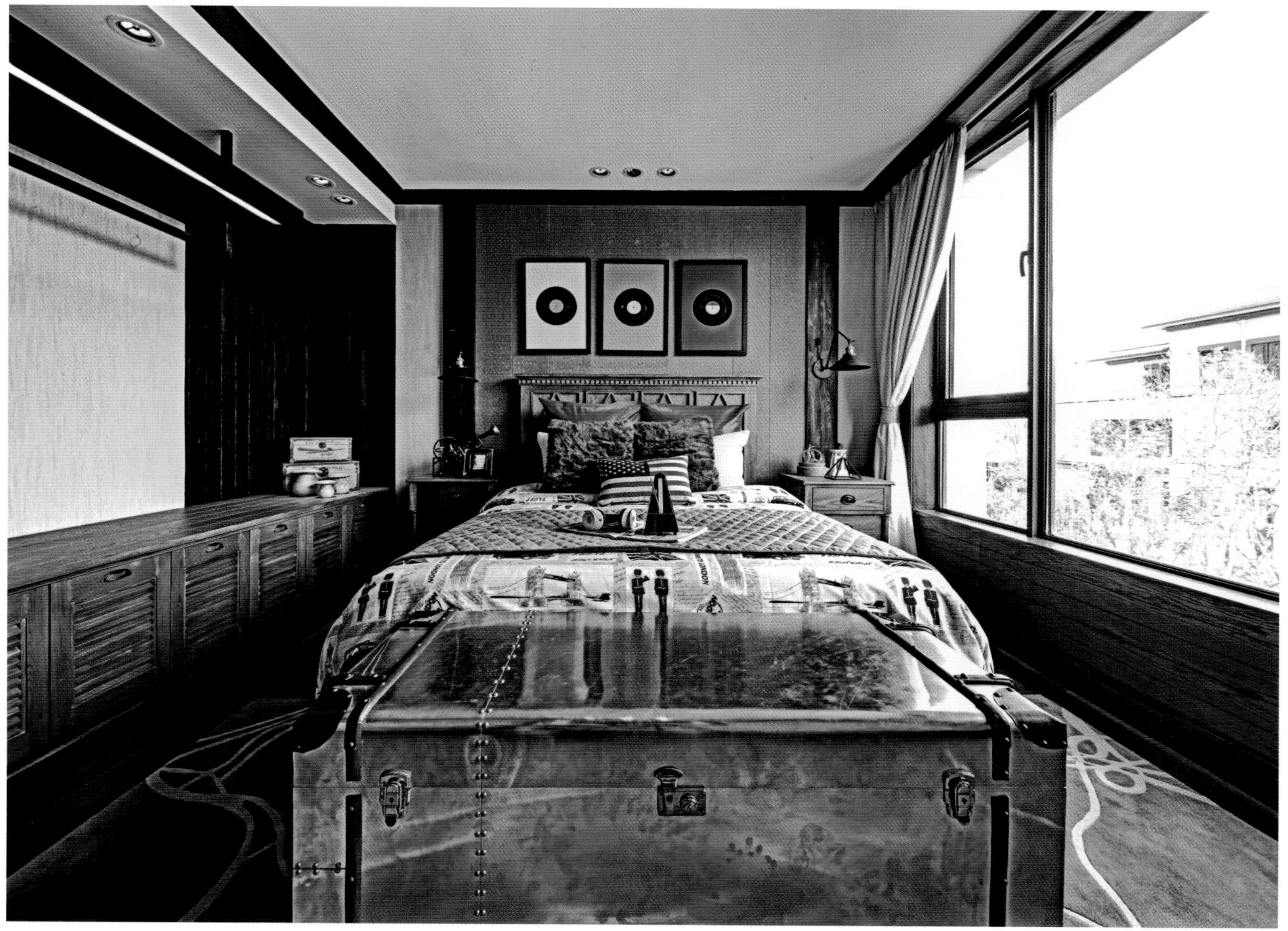

WOOD COLOR BRINGS FORTH FRAGRANCE

木色生香

Design Concept | 设计理念

The biggest visual characteristic of this project is customized dark brown wood furniture all over the house. From hard decoration to soft decoration, from the living room to bedrooms, all emit wood warm fragrant flavors. The living room uses round iron warm color droplight to create focus for the space; large volume European beige sofa manifests formal style of the living room. Light beige wallpaper and floral carpet add free and leisure meanings for the formal space. In the American open kitchen, the intervention of the bar symbolically separates kitchen and dining room; when guests come, they can cook with the owners, which not only promotes their relations, but also makes them feel the pleasure of participation. Round dining table and round droplight echo up and down; two round archaize wall mirrors and a baroque mirror match with four pieces of European furnishing paintings, instantly promoting the artistic atmosphere of the living room.

本案呈现出的最大视觉特色便是全屋定制的深褐色木质家具，从硬装到软装，从客厅到卧室都散发出木质温暖生香的气息。客厅以圆形铁艺暖色吊灯为空间营造聚焦点，大体量的欧式米色沙发彰显出会客厅的正式气派。尽管如此，浅色米黄的墙纸和花式地毯也为正式的空间增添了几份自由休闲意味。美式开放式的厨房，吧台的介入不仅象征性地分隔厨房和餐厅，当有客人来时，可以在这里和主人一起动手烹饪美食，既增进主客之间的感情，又让客人能够感受参与的乐趣。圆形的餐桌和圆形的吊灯上下呼应，两面圆形仿古挂镜和一面巴洛克艺术风格挂镜搭配四幅欧式摆件挂画，瞬间提升了餐厅的艺术氛围。

项目名称：大连恒大潭溪郡23
设计公司：上海全筑建筑装饰集团股份有限公司
软装配套：上海全品室内装饰配套工程有限公司
项目地点：辽宁大连

The master bedroom uses white tic-tac-toe and French window to create more romantic and elegant atmosphere for the bedroom. White beddings collocate with colored bed background, which can switch heavily between fresh and elegant, heavy and graceful. The elder's room is mainly dark red, which fits the owner's tranquil mind after experiencing time and tide. A telescope in the children's room manifests the young owner's taste; creamy white beddings intersperse with jumping blue, adding fun and flexibility for the room.

以白色井字格和落地窗为主卧室创造了更多浪漫典雅的氛围，白色床品搭配黑色床头背景，可在清新典雅和厚重文雅中自由切换；长者房以深红为主色调，契合屋主历经岁月后归于宁静的心境；儿童房一架望远镜就彰显出小主人的品位，米白色的床品点缀跳跃蓝色，更增加了儿童房的趣味性和灵动性。

REAPPEARING AMERICAN ELEGANT LIFE

再现美式优雅生活

Design Concept | 设计理念

This project uses diversified space structure, color and soft decoration to combine nostalgic romantic feelings with modern people's life demands, luxurious and elegant, fashionable and modern, reflecting individualized aesthetic views and cultural tastes. On space layout, the design of large French window not only endows the interior with good day lighting but also brings the outside beautiful scenery inside at the same time has plentiful sunlight. The whole room is filled with sun lights, accompanied by beautiful scenery outside the window; the entire space is extremely comfortable and warm. The designer carefully uses smooth and exquisite marble and high texture wallpaper as main materials, manifesting the exquisite and elegant aesthetics of the space and conveying peaceful and connotative space charms.

本案采用多元化的空间构造、注重色彩及软装，将怀古的浪漫情怀与现代人对生活的需求相结合，兼容华贵典雅与时尚现代，反映出个性化的美学观点和文化品位。在空间布局上，大型落地窗的设计，不仅使得室内拥有良好的采光条件，而且在享受充足光照的同时也可将室外美景尽揽于眼下。满室盈满阳光，窗外美景相伴，整个空间显得特别舒适、温馨。设计师精心选择光洁细腻的大理石、质感上佳的墙纸等为主要原料，彰显出居室精致优雅之美，传达出平和而富有内涵的空间气韵。

项目名称：宁波金色荣湾样板房
设计公司：古木子月(香港)国际空间策划装饰设计事务
设 计 师：胡荣
项目地点：浙江宁波
项目面积：380㎡
主要材料：护墙板、大理石、墙纸等

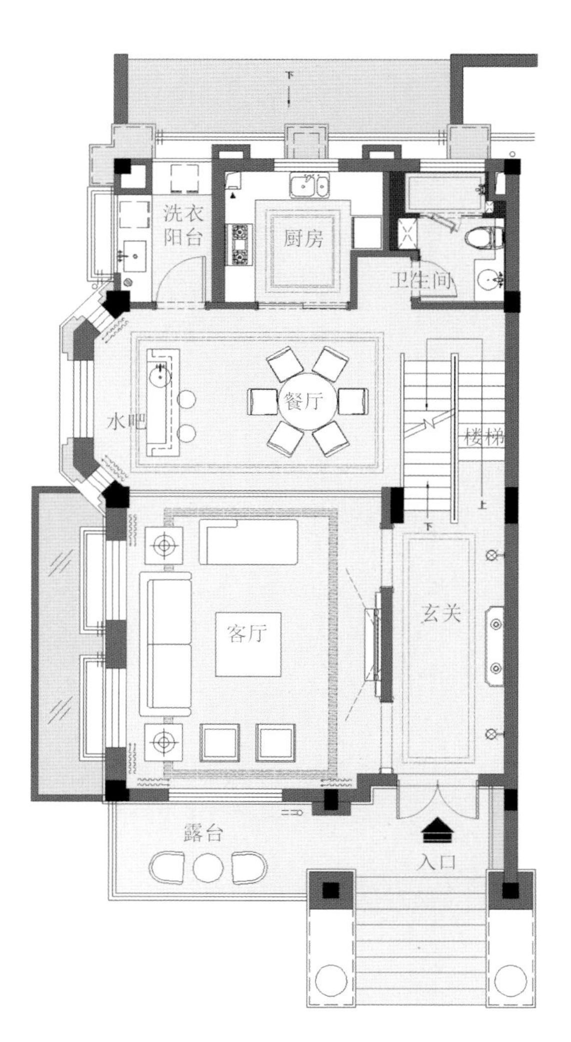
下
洗衣
阳台
厨房
卫生间
水吧
餐厅
楼梯
客厅
玄关
露台
入口

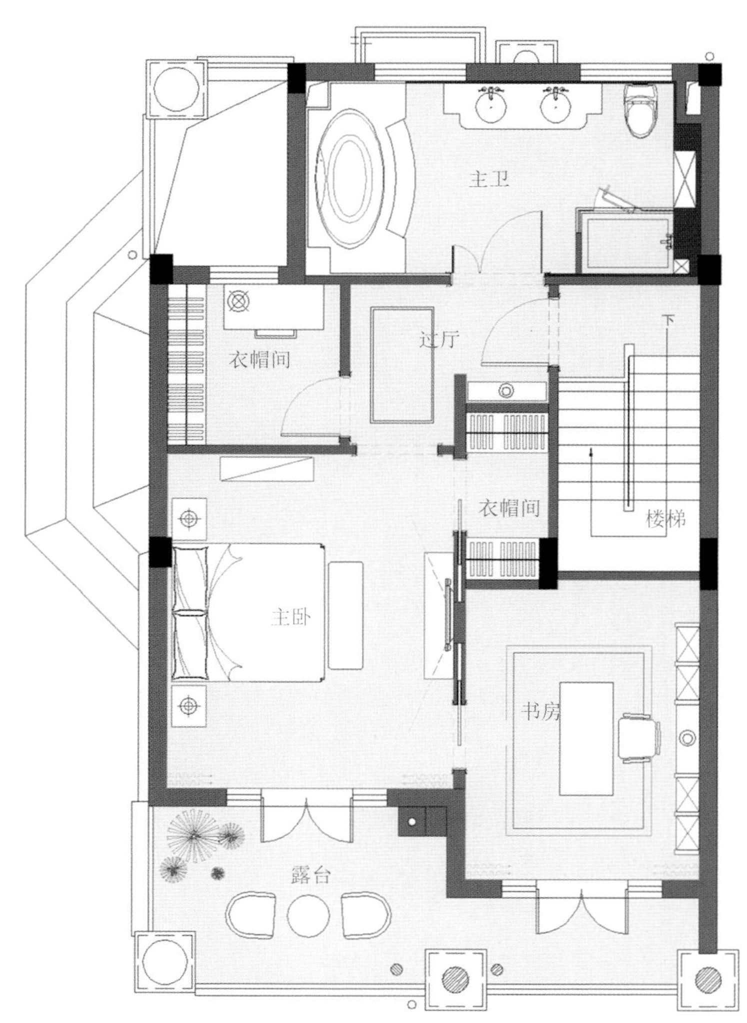
主卫
衣帽间
过厅
下
衣帽间
楼梯
主卧
书房
露台

KNIGHT

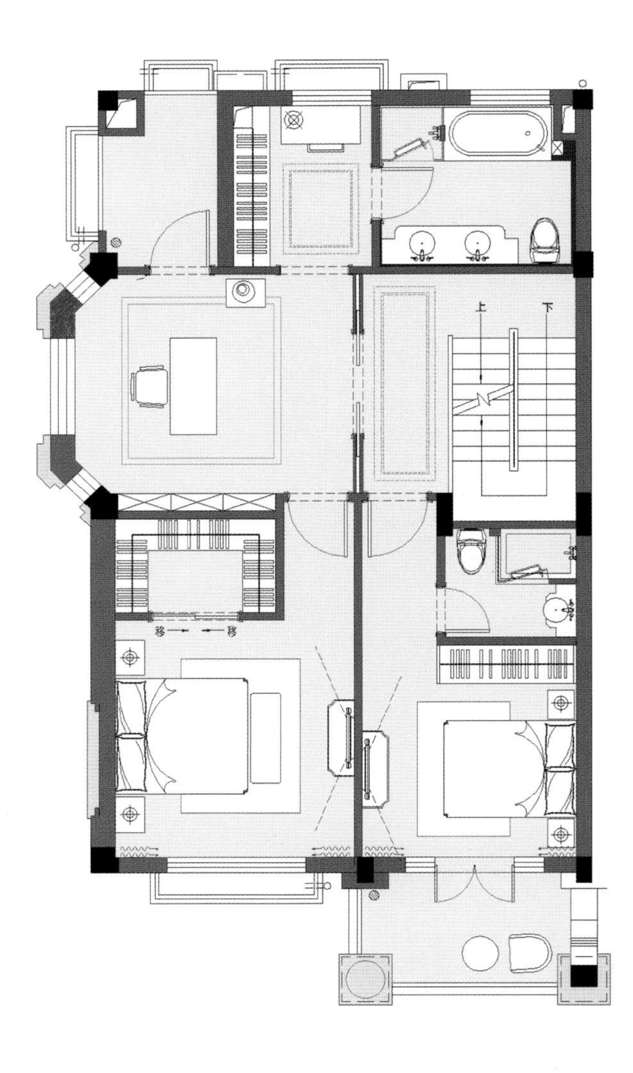

工人房
储物间
楼梯
影音室
通道
红酒吧
台球室
采光井

MOUNTAIN SCENERY WITH BRIGHT SUNSHINE

山语清晖

Design Concept | 设计理念

This project is a single-family villa composed of two duplex villas. The owner's required functions are leisure and vacation. Besides functions featured by common residences, it has enterprise and club functions such as reception, conference and recreation.

To meet requirements of owner, the designers greatly transform the original structure. The first floor is all arranged as spacious spaces with main function as reception; the dining room and kitchen also have reception function. Though they are all spacious spaces, they are well-organized. According to the owner's life and work habits, the kinetonema is organized according to the frequency of utilization and degree of opening progressively to arrange every function area. The second floor is mainly half-open spaces with meeting room and recreational room. The third and forth floors are bedrooms including guests room, fully considering functionality and privacy issues.

本项目是由两套双拼别墅改造而成的独栋别墅，业主要求的功能是休闲、度假，在具有普通住宅应该有的功能以外，还要有接待、会议、洽谈、娱乐等具有企业会所性质的功能。

为达到业主的要求，设计师对原始结构进行了大幅度的改造。一层全部安排为开敞空间，主要功能是接待，包括餐厅和厨房也都具有接待功能。虽然为全开敞的空间，但也要组织得开而不乱。根据业主的生活习惯和工作习惯，组织的动线是按使用频率和开放程度层层递进，以此为依据安排各个功能区。二层为半开放空间，设置了会议室和娱乐室。三层四层为卧室，兼具接待客房，充分考虑了功能性和私密性等问题。

项目名称：戴斯大卫营
设计公司：重庆微分室内设计有限公司
设 计 师：吴钒、梁瑞雪、吴杨武
项目地点：重庆
项目面积：500㎡
主要材料：仿古砖、质感漆、文化石等

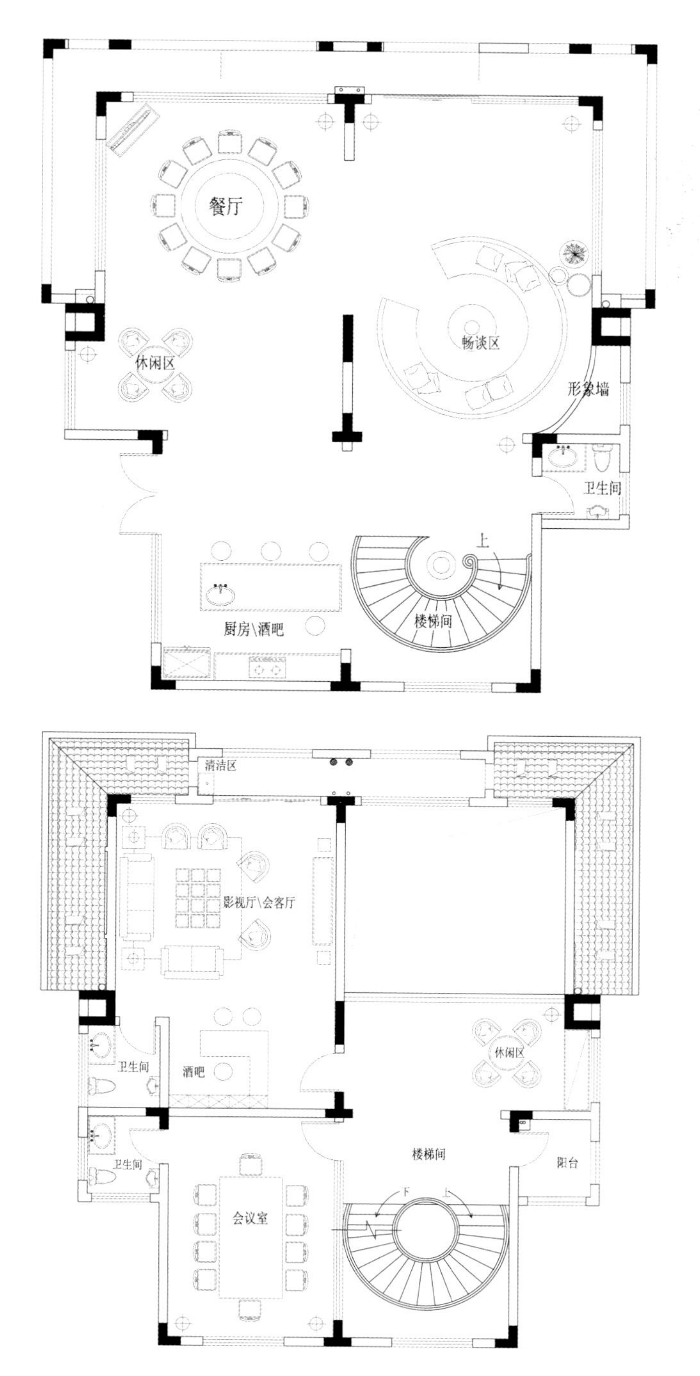

To bring fresh air of the mountain and beautiful natural scenery into the first floor, the designers create many large windows on the premise of harmonious façade. The window type designs make the interior style and outside façade unified. They replace the original small narrow stair by big steel spiral staircase, manifesting modern flavors. To further improve sense of space, they remove the living room ceiling to make it a higher space and bend wood beams to become arced roof, giving the top of the high living room a visual focus.

The style tends to be relaxing, casual, fresh and natural at the same time hale and masculine. The hard decoration stresses openness and inclusiveness, making easiness and leisure, firmness and masculinity harmoniously coexist. The soft decoration chooses products with light loft style; rough and heavy materials such as steel, old wood, imitated stone and rivet leather become the mainstream, collocated with other soft and warm, transparent and lightsome materials, making the space feelings not too monotonous. Lamps and lanterns combine plain and clumsy iron and lightsome and clear glass; iron cushion matches with warm velvet carpet; old woods are easy to become dull so that it is necessary to add some cool iron and glass. Fresh and modern glass utensils are used as flower bottles; dead branches picking from outside endow the big environment a slight of freshness.

为了把一楼与山区清新的空气、美丽的自然风光融合起来，在顾及外立面协调的前提下，尽量多开窗、开大窗，窗型设计兼顾了室内风格与外立面的协调统一。换掉原来狭窄逼仄的小楼梯，用钢结构制作大旋转楼梯，现代气息立显。为进一步提升空间感，打掉客厅天花板使之成为挑高空间。并且用木梁弯曲成拱形屋架，使挑高的客厅顶面有视觉焦点。

风格定位上倾向于轻松、随意、清新自然，同时不失硬朗和阳刚。硬装上讲究开放和包容，让轻松休闲、坚毅阳刚能在其中和谐共存。软装选择的是带有轻loft风格的产品，钢铁、做旧木材、仿石材、铆钉皮革等粗犷厚重的材质成为主流，但同时又要考虑到其他或柔软温暖或通透轻盈的材质与之搭配，使空间感觉不至于太单一。灯具是质朴拙笨的铸铁铁艺与轻盈灵动的玻璃的结合；与铁艺坐具搭配的是温暖的绒毛地毯；做旧木材容易显得沉闷，需要加入少量的铁艺或玻璃的清冽；清新现代的玻璃器皿作为花器，插上捡来的室外掉落的枯枝，给粗犷的大环境带来一丝小清新。

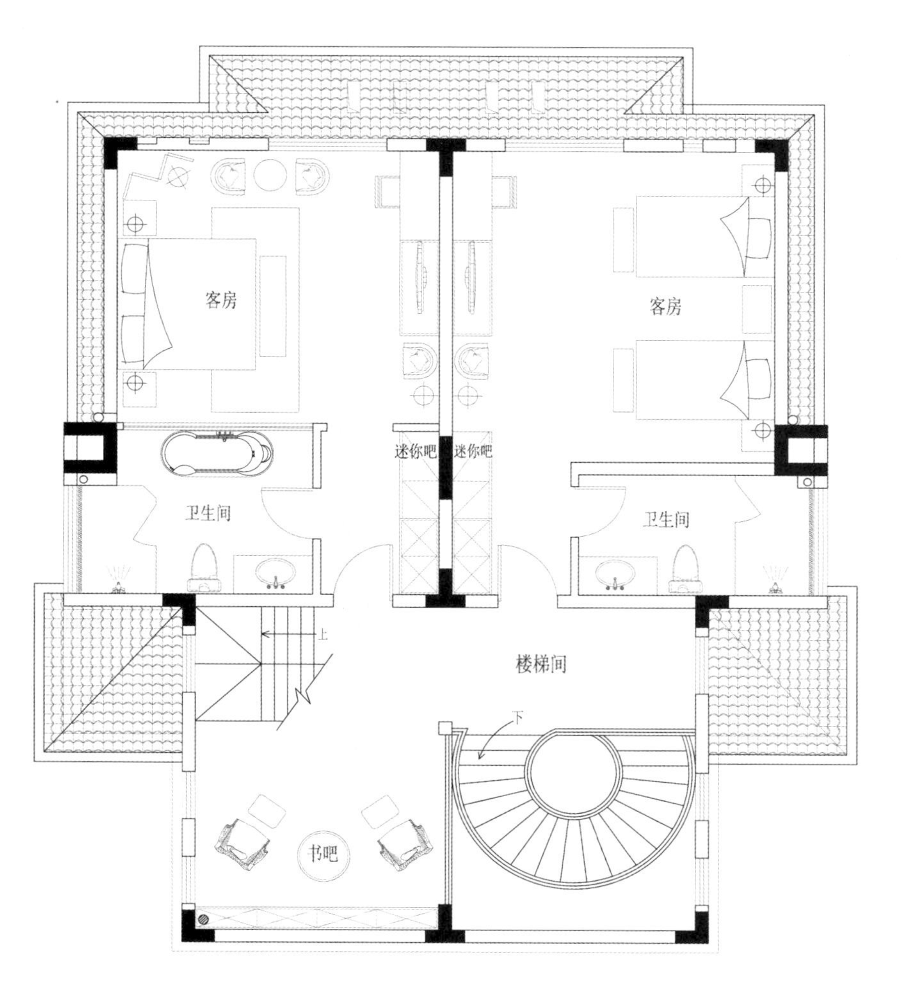
客房
客房
迷你吧
迷你吧
卫生间
卫生间
上
楼梯间
下
书吧

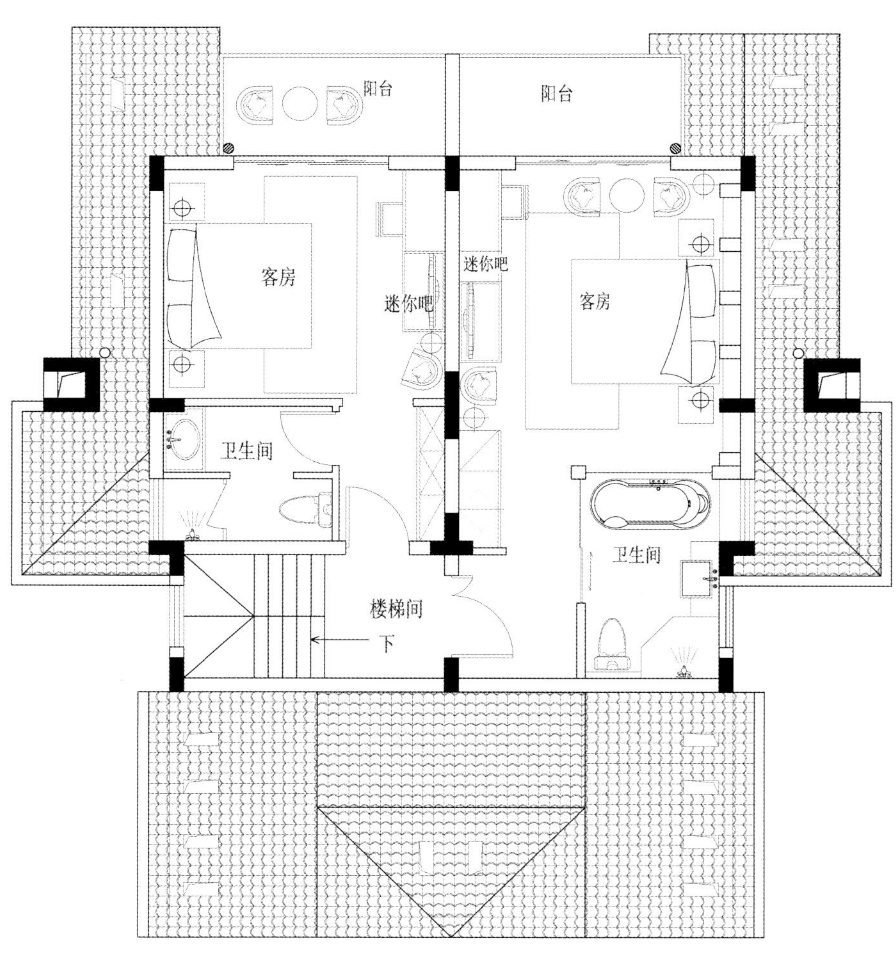
阳台
阳台
客房
迷你吧
迷你吧
客房
卫生间
卫生间
楼梯间
下

TIME COLLOCATION

岁月典藏

Design Concept | 设计理念

The space transformation is the focus of this project. In the original structure, the stair is in the corner; the three floors are three flats without dialogues between the upper and lower space. The train of thought of reforming the layout is to break this pattern and restore the features of the villa. The spiral staircase is newly created in the central of the house, connecting the three floors. All functional areas are set up round the stair, activating the spaces. At the same time, it naturally divides the dining and kitchen area with the static area, solving the problem that the dining room and kitchen cannot be moved and must be as the same floor with bedrooms. You can imagine that it is a big house under a century-old tree and emits the smell of annual ring. The designer uses a lot of primitive materials such as charcoal grilled fir, culture stone and diatom ooze to create a kind of soft and texture living atmosphere.

空间改造，是本案的重点。原始结构中，楼梯偏窝一角，三层也就三大平层，上下空间没有对话。重整布局的思路是打破这格局，还原别墅的特色。在屋中央，开洞新铸了一把旋转楼梯，连通三层。而各个功能区，全围着这把楼梯分布，盘活了空间。同时把餐厨区与静区自然分隔出来，解决因餐厨无法移动，一定要和睡房同层的问题。想象它是个百年老树环绕下的大屋，散发着年轮的味道。设计师用了大量的炭烤杉木、文化石、硅藻泥等古朴的材质，来打造这种柔和、有质感的家居氛围。

项目名称：福州橡树湾

设计公司：新谛室内空间营造社

设 计 师：卓新谛、卓友彬

项目地点：福建福州

项目面积：420㎡

主要材料：炭烤杉木、复古砖、文化石、硅藻泥、墙纸、金钢板等

GOOD NIGHT

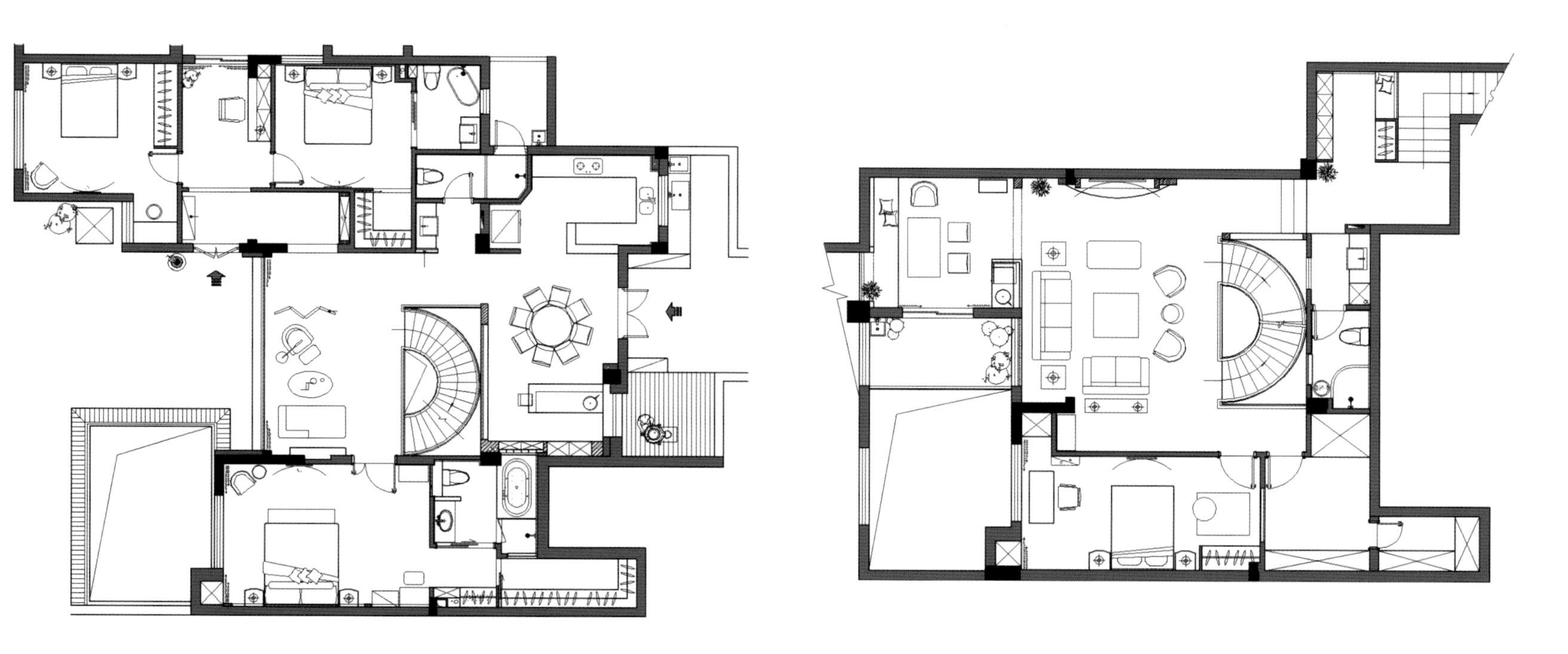

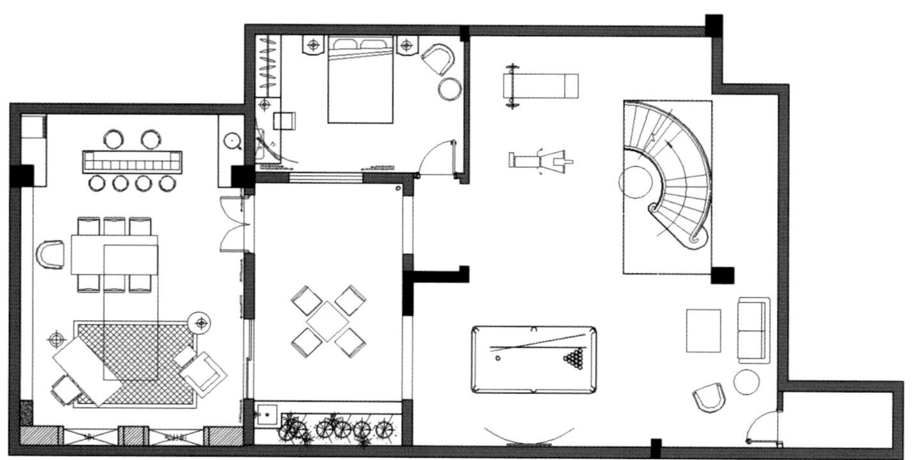

图书在版编目（CIP）数据

轻豪宅风范. II，奢而不侈 豪而不夸 / 深圳视界文化传播有限公司编. -- 北京 ：中国林业出版社，2017.5
ISBN 978-7-5038-8995-0

Ⅰ. ①轻… Ⅱ. ①深… Ⅲ. ①住宅－室内装饰设计 Ⅳ. ①TU241

中国版本图书馆CIP数据核字（2017）第096030号

编委会成员名单
策划制作：深圳视界文化传播有限公司（www.dvip-sz.com）
总 策 划：万绍东
编　　辑：杨珍琼
装帧设计：潘如清
联系电话：0755-82834960

中国林业出版社 • 建筑分社
策　　划：纪　亮
责任编辑：纪　亮 王思源

出版：中国林业出版社
（100009 北京西城区德内大街刘海胡同 7 号）
http://lycb.forestry.gov.cn/
电话：（010）8314 3518
发行：中国林业出版社
印刷：深圳市雅仕达印务有限公司
版次：2017年5月第1版
印次：2017年5月第1次
开本：235mm×335mm，1/16
印张：20
字数：300千字
定价：428.00元（USD 86.00）